Fifty Machines

That Changed the Course of

History

A FIREFLY BOOK

Published by Firefly Books Ltd. 2012

First printing

Publisher Cataloging-in-Publication Data (U.S.)
Chaline, Eric.
Fifty machines that changed the world / Eric Chaline.
[224] p. : ill. ; cm.
Includes bibliographical references and index.
ISBN-13: 978-1-77085-090-3
1. Inventions — History. 2. Machinery — History. I. 50 machines that changed
the world. II. Title.
609 dc23   T15.C343 2012

Library and Archives Canada Cataloguing in Publication
Chaline, Eric
Fifty machines that changed the world / Eric Chaline.
Includes bibliographical references and index.
ISBN 978-1-77085-090-3
1. Machinery—History. 2. Machinery—Social aspects.
I. Title.
TJ15.C43 2012   621.809   C2012-901426-5

Published in the United States by
Firefly Books (U.S.) Inc.
P.O. Box 1338, Ellicott Station
Buffalo, New York 14205

Published in Canada by
Firefly Books Ltd.
66 Leek Crescent
Richmond Hill, Ontario L4B 1H1

Conceived, designed and produced by
Quid Publishing
Level 4, Sheridan House
114 Western Road
Hove BN3 1DD
England

Cover and interior design: Lindsey Johns

Printed in China

Fifty Machines

That Changed the Course of

History

written by Eric Chaline

FIREFLY BOOKS

CONTENTS

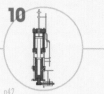

50

MACHINES
THAT
CHANGED
THE COURSE
OF HISTORY

Humanity has had a complex and often contradictory relationship with the machine. The development of each new technology has brought about unforeseen transformations in society, politics, economics and the natural environment, ending, sometimes in a matter of years, ways of life that had endured for centuries. Humans may like to believe that machines are their servants, but as we shall see in this survey of the iconic machines of the past two centuries, they have often been the true masters that have unmade and remade human lives, livelihoods and lifestyles.

The handy man

Members of the *Homo habilis* (2.3–1.4 million years BP) species of hominids, who figure as a major branch in humanity's flourishing ancestral tree, distinguished themselves from their predecessors by their superior ability to make and use stone tools, earning the name "handy man," and setting humanity on the road that would one day lead to the steam locomotive, the vacuum cleaner, the PC, and the Hubble telescope. This survey of iconic machines that changed the course of history does not go back to the invention of the hand ax or wheel, but begins in 1801, with the first successful application of automation to weaving, which had until then been the preserve of the skilled artisan. From that time on, for better, or for many of those directly involved at the time, for worse, machines have continued their relentless advance into every aspect of human life and culture.

"A tool is but the extension of a man's hand, and a machine is but a complex tool. And he that invents a machine augments the power of a man and the well-being of mankind." HENRY WARD BEECHER (1813–1887)

During the First Industrial Revolution (1760–1860), machines revolutionized tool-making (Roberts' lathe, page 12; Whitworth's planing machine, page 24) and the manufacture of consumer goods, particularly textiles (Jacquard loom, page 8; Roberts' loom, page 20; Corliss steam engine, page 26), whose production became mechanized and automated, turning the skilled artisan into the unskilled factory worker, and transport, with the development of steam locomotives and steamships (Rocket, page 14; SS *Great Eastern*, page 38).

The Second Industrial Revolution (1860–1914) provided a new source of power: electricity (Gramme machine, page 44; Parsons' steam turbine, page 52; Westinghouse AC, page 58) and saw an even greater transformation of society as technology entered the office (Linotype, page 46; Underwood No. 1 typewriter, page 84; Tungsram light bulb, page 94; Automatic Electric candlestick telephone, page 98) and the home (Singer "Turtle Back" sewing machine, page 36; Hoover suction sweeper, page 110), and revolutionized transport ("Rover" bicycle, page 54; Diesel engine, page 78; Model T Ford, page 104) and popular entertainment (Berliner "Gramophone," page 60; Lumière "Cinématographe," page 66; Marconi radio, page 72).

The manufacturing of the modern world

In what is often called contemporary "post-industrial" society—rather oddly, I've always thought, as a manufacturing industry and technology are still the necessary foundations of all human societies—machines have relieved humans of the most repetitive tasks in the factory ("Unimate" industrial robot, page 170), home (G.E. top-loading washing machine, page 146; Victa lawn mower, page 160), and office (IBM PC 5150, page 200); and given humans new ways to fill their increasing leisure time (Baird "Televisor," page 130; Ampex reel-to-reel tape recorder, page 150; JVC HR-3300EK VCR, page 184; Atari 2600, page 188; Sony "Walkman," page 192), to communicate (Hayes Smartmodem, page 206; Motorola "StarTAC" cell phone, page 214), to produce energy (Magnox nuclear reactor, page 164; Vestas wind turbine, page 196), and to travel (LZ 127 Graf Zeppelin, page 126, De Havilland Comet jetliner, page 154). Last but not least, machines have allowed humans to investigate and explore their universe in ways undreamed of by their ancestors (Siemens electron microscope, page 138; EMI CT scanner, page 180; Saturn V rocket, page 174; Hubble space telescope, page 210).

01

Designer:

Joseph Marie Jacquard

JACQUARD LOOM

Manufacturer:

Joseph Marie Jacquard

Industry

Agriculture

Media

Transport

Science

Computing

Energy

Home

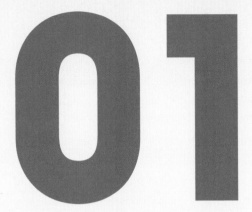

1801

Until the Industrial Revolution, weaving was a highly labor-intensive process—in particular for luxury silk brocades, which were made by closed guilds of artisans who owned and operated their own looms. Although not the first attempt at automation in silk weaving in France, Jacquard's loom was the first to succeed commercially.

Joseph Marie Jacquard

The emperor's inventor

The Jacquard loom is a good place to start on our survey of the 50 machines that changed the world because, like most of the devices featured in this book, it is not an invention that was created by a single inspired individual, but the development of a much older device—the treadle loom—building on and incorporating the improvements of previous innovators. Joseph Marie Charles (1752–1834), whose branch of the Charles family had been given the nickname "Jacquard," wanted to automate the laborious process of silk weaving, in order to eliminate human error in the production of complex patterned textiles, and at the same time to free child laborers, who were employed as "draw boys" in silk weaving on the traditional two-man drawloom. In the former aim, he had the sponsorship and encouragement of the future Emperor Napoleon I (1769–1821), who wanted to challenge English dominance of the textile industry.

The Jacquard loom became the basis for a worldwide silk industry. Shown here is silk brocade manufacture in South Manchester, CT, c. 1914.

DEVELOPMENT OF THE LOOM

Back-strap loom	**Neolithic period**
Warp-weighted loom	**Neolithic period**
Foot-treadle loom	**c. 300** CE
Bouchon loom	**1725**
Falcon loom	**1728**
Vaucanson loom	**1745**
Jacquard loom	**1801**

A native of Lyons, the capital of the French silk industry and the scion of a silk weaver's family at the dawn of the French Industrial Revolution, Jacquard was in just the right time and place to develop his invention. Jacquard's first attempt was displayed in 1801 at the Industrial Exhibition in Paris, but it was only when he saw the loom made over half a century earlier by Jacques Vaucanson (1709–1782), which used primitive punch cards, that he was able to perfect his own invention. Napoleon, by now Emperor of France, inspected the finished loom in 1805, and granted Jacquard a lifetime pension. By 1812, the year of the emperor's ill-fated Russian campaign, there were 11,000 Jacquard looms in France, despite the fierce opposition of the silk weavers who feared the machine would ruin their livelihoods. Although Jacquard had succeeded in his aim of automating silk weaving and thus introduced automation into industrial production, he failed to improve the lot of draw boys, who had to find more dangerous jobs in mills and factories once they were no longer needed in silk weaving.

"Jacquard's invention has produced a comprehensive revolution in manufacturing; it has set a dividing line between the past and the future; it has initiated a new era in progress."

FRANÇOIS MARIE DE FORTIS, *IN PRAISE OF JACQUARD* (1840)

THE JACQUARD LOOM

[A] Jacquard mechanism
[B] Punch cards
[C] Comb
[D] Warp
[E] Thread spool
[F] Frames
[G] Treadle

KEY FEATURE:
THE PUNCH CARD

Jacquard did not invent the punch card, which was first devised by Jean Falcon for his loom of 1728, but he improved the mechanism for his 1805 design. The pattern of holes on the cards is a basic "program" that creates the final design. A posthumous portrait of Jacquard, woven on a Jacquard loom in 1839, required 24,000 punch cards. Jacquard's use of punch cards is also seen as a key development in the history of computing.

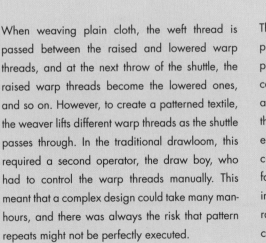

[F]

[C]

[D]

[B]

[A]

[G]

[E]

When weaving plain cloth, the weft thread is passed between the raised and lowered warp threads, and at the next throw of the shuttle, the raised warp threads become the lowered ones, and so on. However, to create a patterned textile, the weaver lifts different warp threads as the shuttle passes through. In the traditional drawloom, this required a second operator, the draw boy, who had to control the warp threads manually. This meant that a complex design could take many man-hours, and there was always the risk that pattern repeats might not be perfectly executed.

The basis of the Jacquard mechanism was the punch card that carried the information needed to produce the design. Hooked rods "read" the punch cards as they came through the mechanism around a perforated square prism. Depending on whether the rods encountered a hole or solid paper, they either raised or lowered the warp threads, thereby creating the design perfectly each time, and much faster than in the traditional method. Jacquard also increased the capacity of the machine, by incorporating eight rows of needles. As each hook could be connected to several threads, the pattern could be repeated more than once.

02

Designer:
Richard Roberts

ROBERTS' LATHE

Industry

Agriculture

Media

Transport

Science

Computing

Energy

Home

Manufacturer:

Richard Roberts

"The hammer, the file, and the chisel did all that was done.
Mr. Roberts soon saw that without better tools than these,
mechanical accuracy could not be obtained."

THE MECHANICS' MAGAZINE (1864)

1817

Richard Roberts was a talented inventor in his own right, who devised the Roberts' self-acting mule, the Roberts' loom (see page 20), and the first gas meter, but he began his career in the design and improvement of machine tools: notably a planing machine and the subject of this entry, the Roberts' lathe.

Richard Roberts

Turn of the screw

When drawing up a list of the most important machines of the nineteenth century, we might think of steam engines, locomotives and power looms, but none of these machines would have been manufactured in any number without machine tools that not only speeded up work and saved on labor and expense, but also ensured much greater accuracy in the manufacture of standardized components. Prior to the nineteenth century, the planing, milling and drilling of wood and metal were achieved with a primitive toolset that had changed little since the Middle Ages. Spurred by the demands of the Industrial Age, British inventors developed a range of high-precision machine tools. Among these, one of the most gifted and prolific was Richard Roberts (1789–1864), who was described by a contemporary as "one of the true pioneers of modern mechanical mechanism."

A lathe is a tool designed to hold a wooden or metal workpiece in place and rotate it so it can be cut, sanded, drilled or shaped. Typical items that are made on a lathe are wooden table legs and metal camshafts. Roberts designed his lathe of 1817 to manufacture metal components with a much higher degree of precision than was possible with traditional methods. The heavy-duty Roberts' lathe was 6 feet (1.8 m) long and made entirely of cast iron, which ensured that it would be accurate even when working with heavy metal components. It consists of a bed with a sliding screw-driven carriage to carry the toolpost, a headstock and tailstock to hold the workpiece in place, a pulley for a belt drive for an external power source (in this case, Mrs. Roberts turning a wheel in the basement) and a back-gear to vary the lathe's working speed. Although not as impressive as our next entry, Stephenson's Rocket, the Roberts' lathe probably played a much more important role in the history of engineering and industry by providing the standardized components that made mass production possible.

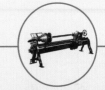

03

Designer:
George Stephenson

STEPHENSON'S "ROCKET"

Industry

Agriculture

Media

Transport ■

Science

Computing

Energy

Home

Manufacturer:

Robert Stephenson & Co.

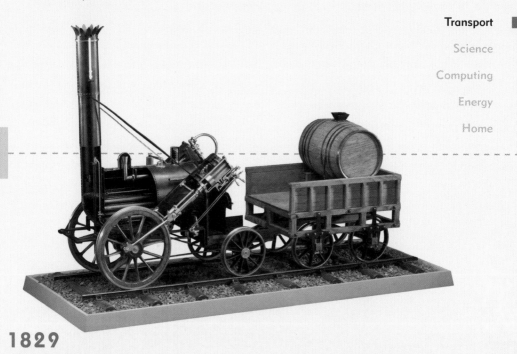

1829

Stephenson's Rocket was not the first locomotive, but it was undoubtedly the most iconic among early steam engines. It was the outright winner of the Rainhill Trials (1829) that decided which locomotive would haul the carriages on the Liverpool to Manchester Railway, the first inter-city railroad for freight and passenger transport to be operated in Great Britain that opened in 1830, 5 years after the world's first public railroad, the Stockton and Darlington of 1825.

Stephenson's trial and triumph

October 1829 witnessed one of the most momentous events of the early railroad age: the Rainhill Trials, whose prize was the contract to build the locomotives that would haul passengers and freight on Britain's first inter-city railroad, the Liverpool and Manchester Railway (L&MR). In the age of the jetliner, the equivalent would be a grand contest between a Boeing 747-400 and an Airbus A380 on the London to New York route. The trials took place on a section of track near the village of Rainhill, 9.3 miles (15 km) east of the port of Liverpool, one of the great urban centers to emerge during Britain's First Industrial Revolution.

The trials, held between October 7 and 14, attracted huge public interest, and the attendance of dignitaries, crowds of sightseers and representatives of the nation's press. In addition to the £500 prize, the victor would not only win the contract for the new railroad but also national and international fame, and no doubt lucrative contracts for many years to come. Among the five entrants, which included a throwback to an earlier age, the horse-powered Cycloped, the main contenders were the Rocket, designed by George Stephenson (1781–1848) and his son Robert (1803–1859), the Sans Pareil and the Novelty. The rules of the competition specified the maximum weight and design features of the engine, and the trials themselves consisted of traveling over a set distance of 35 miles (56 km), both with and without a load, to establish the engine's average speed and fuel consumption.

The Novelty　　　　　The Sans Pareil　　　　　The Rocket

In the 1830 modification of the Rocket design, the piston and cylinders were set horizontally, improving the engine's overall stability. This became standard on later locomotives.

The Sans Pareil was almost disqualified at the start for being over the maximum weight but it matched the Rocket's performance until it cracked a cylinder and was forced to retire. The real competition came from the Novelty, which astonished the public by reaching the unheard-of speed of 32 mph (51.5 km/h) while hauling a fully laden carriage. However, it, too, was forced to retire because of technical difficulties. The Rocket was the only locomotive to complete the trials, averaging 12.5 mph (20 km/h) while hauling a load three times the weight of the engine, and 24 mph (38.5 km/h), with a carriage and passengers.

"The 'Rocket' showed that a new power had been born into the world, full of activity and strength, with boundless capability of work. It was the simple but admirable contrivance of the steam blast, and its combination with the multitubular boiler, that at once gave locomotion a vigorous life, and secured the triumph of the railway system." SAMUEL SMILES, THE LIFE OF GEORGE STEPHENSON (1860)

The self-made engineer

George Stephenson, who is hailed as the "Father of the Railways," was born in a time when there were no technical schools, polytechnics or universities that taught science and mechanical engineering. He began life in the most humble of circumstances, the son of a collier. He had to start working as a child, receiving little formal education. Aged 14, he obtained a job as a brakesman on a tramway and then as a stoker on an engine at a coal mine. Principally self-taught, he began to carry out repairs on the engine he tended. He married in 1803, fathering his only son and future business partner, Robert. In 1816, having demonstrated his technical abilities, Stephenson built his first engine, the Blücher (named for the Prussian general who had helped the Duke of Wellington [1769–1852] defeat Napoleon at Waterloo the year before), which like most locomotives at that time was used to haul coal in a colliery.

From building locomotives, Stephenson took on the design of entire railroads—then still only a few miles long—including bridges and tunnels, as well as a new design for cast-iron rails, which replaced the earlier wooden rails that were no longer able to withstand the weight of heavier engines. Stephenson established his national reputation when he planned and built the 26-mile (42-km) Stockton and Darlington Railway (S&DR), which opened in 1825, and was the first in the world to carry both fare-paying passengers and freight. Stephenson built several engines for the S&DR, including the Locomotion. On the inaugural journey, Locomotion hauled 600 passengers at what we would consider the snail-pace speed of 10–12 mph (16–19 km/h) but which amazed a public used to traveling either on foot or in slower and much bumpier stagecoaches. The age of the train had begun.

Hailed as the "Father of the Railways," George Stephenson left school at 14 and was entirely self-taught.

1781–1848

George Stephenson

17

First rail passenger tragedy

Stephenson's victory established his reputation as the world's leading manufacturer of locomotives. The L&MR, however, was almost never built. When George Stephenson entrusted the surveying of the route to an employee, his son Robert having gone to South America, the whole project was almost thrown out by Parliament in 1825, because the survey was shown to be inaccurate. In addition, the railroad faced fierce opposition from landowners who did not want it to pass through their land, and rivals, such as the canals, which until then had carried the heavy freight. Stephenson was dismissed but then reinstated a year later. The 35-mile (56-km) line, completed in 1830, had many firsts: the first tunnel under an urban area, a cutting 70 ft (21 m) deep, 64 bridges and viaducts, and a 4.75-mile (7.5-km) section that was "floated" on the surface of Chat Moss peat bog.

The day of the line's inauguration attracted dignitaries, including the Prime Minister, the Duke of Wellington. During a break in the proceedings, William Huskisson (1770–1830), a Member of Parliament, alighted from the train and walked along the track to speak to the Duke. Tragically, he failed to notice the Rocket speeding along the other track toward him. Panicking, he fell onto the track, where the locomotive crushed his leg. He died of his injuries a few hours later, despite having been evacuated by train for medical assistance, becoming simultaneously the first injured person to be carried by train and the first passenger railroad casualty.

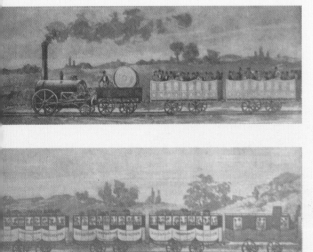

STEAM LOCOMOTIVES

Murdoch's steam engine	1784
Trevithick's Pen-y-darren locomotive	1804
Murray's Salamanca	1812
Stephenson's Blücher	1816
Stephenson's Locomotion	1825
Hackworth's Royal George	1827
Stephenson's Rocket	1829

The Rocket hauled passengers and freight on the world's first "inter-city" railroad line.

STEPHENSON'S "ROCKET"

Unlike later locomotives that ran on six wheels to distribute their weight, Rocket had a 0–2–2 wheel arrangement with a separate tender for coal and water. The steam was provided by a furnace 2 x 3 ft (61 x 91 cm) that heated the multitubular boiler 6 x 3 ft (1.8 x 91 cm). Rocket had two cylinders set at a 35 degree angle, with the piston driving the pair of front driving wheels 4 ft 8 in (1.46 m) in diameter with connecting rods. The rear wheels, which were uncoupled from the driving wheels, were 2 ft 6 in (79 cm) in diameter. In earlier engines, pistons were often set vertically, which caused the engine to sway. The angle of the piston improved the stability of the engine, and in the 1830 modification of the Rocket, the cylinders were set horizontally, which soon became standard on later locomotives. Rocket had two safety valves and a blast-pipe that fed exhaust steam from the cylinders into the base of the chimney to create a partial vacuum and pull air through the fire.

[A] Furnace
[B] Multitubular boiler
[C] Chimney
[D] Inclined cylinders
[E] Connecting rods
[F] Safety valves
[G] Exhaust steam pipe
[H] Fuel tender
[I] Water cask

KEY FEATURE:
THE MULTITUBULAR BOILER

One of the major advances of the Rocket was its multitubular boiler, with 25 3-inch (7.6-cm) diameter copper tubes running the length of the boiler instead of a single pipe. This greatly increased the surface area for heat transfer, making the engine far more efficient and powerful enough to haul heavy loads. Later boiler designs would increase the number of tubes.

04

ROBERTS' LOOM

Industry ■

Agriculture
Media
Transport
Science
Computing
Energy
Home

Manufacturer:

Sharp, Roberts, and Co.

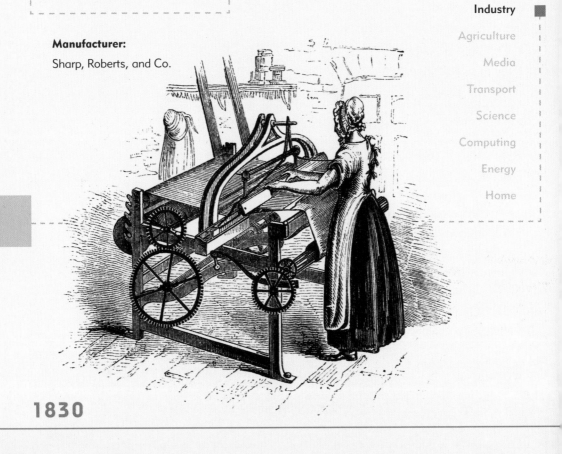

1830

Although the first power loom was built in 1785, it was not until Roberts created his own loom almost half a century later that power looms began to challenge handlooms in the production of cloth in Britain. As was seen in his machine tools, the Roberts' loom was built with both solidity and high precision, made from standardized parts that could be mass-produced.

Weaving pioneer and draft dodger

We first met the talented British engineer and inventor Richard Roberts (1789–1864) when we featured one of his high-precision machine tools, the Roberts' lathe (page 12). In designing and building machinery for the textile industry, Roberts showed the same ingenuity and concern for accuracy and high precision, so that his machines were not individually crafted one-offs but made using standardized components that could be mass-produced. In addition to the power loom, Roberts mechanized spinning with his revolutionary self-acting mule of 1825. Together, these inventions helped transform cloth manufacture from a skilled craft into a mechanized industrial process. His cast-iron looms were both sturdy and reliable, and were adopted in large numbers by the Lancashire mill owners. It would take another 12 years for weaving to become fully automated with the introduction of the Lancashire loom that used many of Roberts' innovations.

The mechanization of weaving dramatically reduced the numbers employed in the industry.

POWER LOOM

Cartwright loom	1785
Radcliffe loom	1802
Horrocks loom	1813
Moody loom	1815
Roberts' loom	1830

Like many of his contemporaries, Roberts came from a humble background and received only the most rudimentary formal education. He was the son of a shoemaker in the village of Llanymynech on the border between Wales and England. After receiving basic schooling from the parish priest, he found a job on a canal and then at a limestone quarry. In his twenties he found work as a pattern maker at an iron works, later making foreman. To avoid being drafted into the militia and sent to fight against Napoleon, he went to Manchester where he found his first job in lathe and tool making. However, still pursued as a draft dodger, he walked to London, where he found work with Henry Maudslay (1771–1831), one of the pioneers of the British machine-tool industry, whose workshop turned out some of the most talented engineers of their generation. After the end of the war in 1815, Roberts returned to Manchester, where he set up his own business. Although gifted with mechanical and engineering skills, Roberts failed as a businessman. Unlike his great contemporary, Joseph Whitworth (see next entry), and despite his many patents and successful inventions, he died in poverty.

"Roberts' improvements in the loom constituted, in their day, a very distinct step forward."

RICHARD MARSDEN, COTTON WEAVING: ITS DEVELOPMENT, PRINCIPLES AND PRACTICES (1895)

THE ROBERTS' LOOM

A loom's hardwood shuttle with thread bobbin.

[A] Cast-iron frame
[B] Batten
[C] Warp beam
[D] Roller
[E] Heddles
[F] Cloth beam
[G] Breast beam

KEY FEATURE:
TAKING-UP GEAR MECHANISM

According to Marsden's *Cotton Weaving* (1895), the most original feature of the Roberts' loom, patented by Roberts in 1822, was the taking-up gear that took up the slack in the cloth to prevent breakage in the yarn threads. This consisted of a toothed wheel on the cloth roller that was actuated by a pinion fixed onto a ratchet wheel to control the speed of the yarn passing through the loom.

A traditional shuttle loom performs four basic operations: shedding, which is the raising of the warp threads by the heddles, so that the shuttle can pass through (which can be controlled automatically by a Jacquard mechanism); picking, which is the passage of the shuttle from one side to the other, and includes the weaving of an edge to the cloth; battening, which presses the yarn against the finished fabric; and taking up, which winds the finished cloth onto a beam and releases the warp yarn from the warp beams. These four operations were automated in Roberts' loom, making it the first successful commercial powered loom.

Until the Roberts loom, machines were made with wooden frames, but Roberts introduced a sturdy cast-iron frame made from standardized components. This not only made production much less costly, but it also simplified maintenance, since when a breakdown occurred a new part did not have to be made to order. One of the greatest challenges for early loom designers was to maintain constant tension throughout the machine to avoid breakages of the warp threads, which Roberts achieved with his new taking-up gear. The loom also had a built-in brake and shut-off mechanism. Two levers on the side of the frame threw the shuttle; as it entered the shuttle-box, it depressed a lever that acted as a brake. However, if the lever was not depressed, probably because the thread had broken and the shuttle had not returned to the box, the loom stopped.

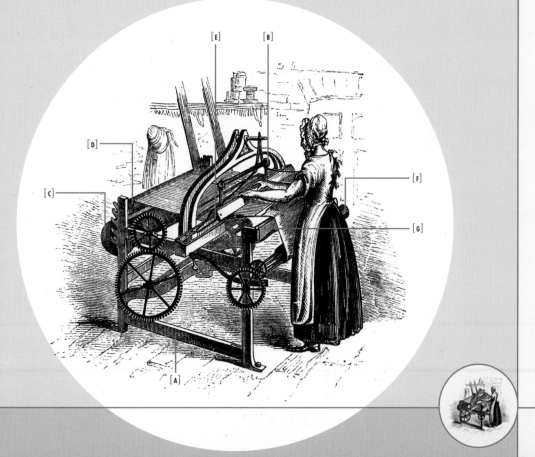

[E] [B] [D] [C] [F] [G] [A]

05

Designer:

Joseph Whitworth

WHITWORTH PLANING MACHINE

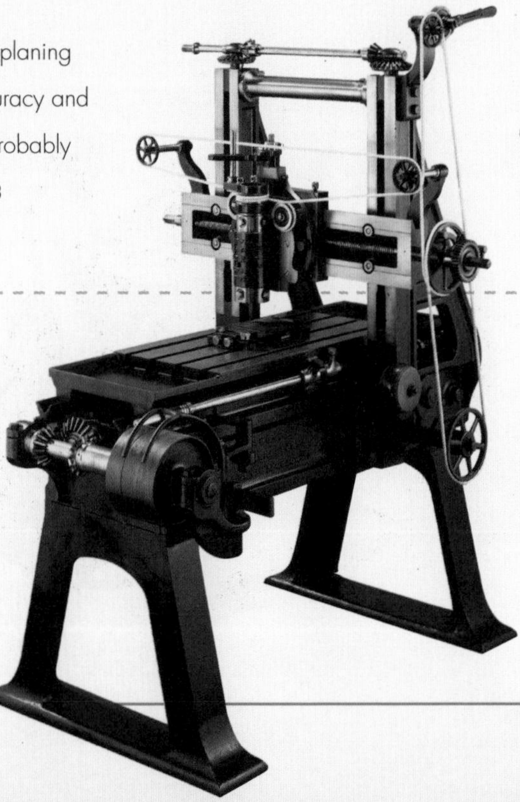

Industry

Agriculture

Media

Transport

Science

Computing

Energy

Home

Manufacturer:

Joseph Whitworth and Co.

"[Mr Whitworth] introduced into the planing machine a degree of structural accuracy and mechanical finish which have not probably been excelled." *The Engineer*, 1863

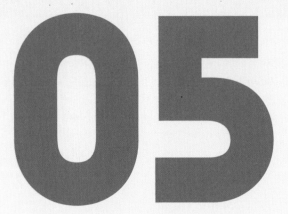

1842

Joseph Whitworth

Like his contemporary and one-time colleague, Richard Roberts, Joseph Whitworth began his career producing high-precision machine tools, including a screw-cutting lathe, and the subject of this entry, the Whitworth planing machine. He is remembered for many achievements, including the first British Standard for screw threads.

"Plane" for the eye to see

Two machines dominate the early machine tool industry: the lathe, an example of which we featured previously, and the subject of this entry, the planer or planing machine, designed by Joseph Whitworth (1803–1887). Hand-planing tools for woodwork have existed since the earliest times, but achieving perfectly flat surfaces in metal is a far more demanding task. Flat metal surfaces were achievable by different methods, but how good they were depended entirely on the skill of the craftsman making the piece. However, even with the keenest eye and lightest of touches, these techniques could not achieve the accuracy and standardization required for mass production, which, added to the labor and time required for handwork, meant that such pieces were prohibitively expensive one-offs.

The exact history of the planing machine is difficult to reconstruct, and various claims have been made for its invention and later improvement. In early designs, the workpiece moved on a table under the suspended cutter. The table moved back and forth in a straight line, allowing the cutter to remove a section of the metal surface, then stepped the cutter to one side to produce an overlapping cut that exactly matched the previous one. Planers were used to manufacture accurate flat surfaces for the components of a wide range of machinery, including steam engines, locomotives and textile machinery. Although Whitworth was undoubtedly not the first to design a planing machine, his powered design was acknowledged to be one of the best, offering greater precision than its predecessors and rivals, while at the same time being simpler to operate.

Unlike Richard Roberts who died a pauper, Whitworth prospered and diversified his machine-tool business into armaments, manufacturing armaments for the British Army to fight the Crimean War (1853–1856). He amassed a considerable fortune, some of which he devoted to the development of technical education, sponsoring the newly established Manchester Mechanics' Institute (now UMIST) and founding the Manchester School of Design.

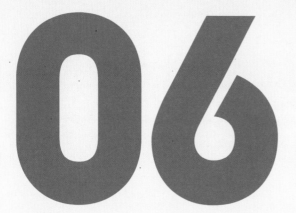

06

Designer:
George Corliss

CORLISS STEAM ENGINE

Industry
Agriculture
Media
Transport
Science
Computing
Energy
Home

Manufacturer:
Corliss, Nightingale, and Co.

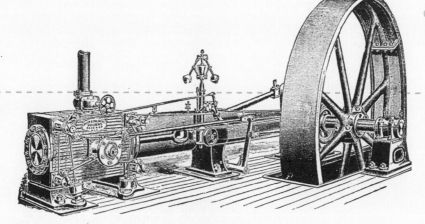

1849

Although the steam engine was an ancient Greek invention, it was not developed for practical use until the end of the seventeenth century. At first supplementary to waterpower, steam eventually became the principal power source of the First Industrial Revolution. The highly efficient Corliss engine completed the process, powering mills and factories, freeing them from the need for a nearby water supply.

George Corliss

America's James Watt

So far in this book, we have featured the inventions of one Frenchman and three Britons. In this latest entry on the Corliss steam engine, we come to one of the first of many iconic North American machines. In discussing the First Industrial Revolution, such is the preponderance of British inventors and engineers that it is easy to forget that technology was being developed in other parts of the world. Since colonial times, the 13 colonies had been developing their own native industries and technologies, and once independent, the United States would quickly catch up and rival Britain's industrial achievements.

Although originally a French invention of 1690, the stationary steam engine was commercialized in England by Thomas Newcomen (1664–1729), and subsequently improved, most famously by James Watt (1736–1819).

Although there were many improvements made to Watt's original design in the next few decades, the man who is considered to be Watt's rightful successor was George Corliss (1817–1888), who patented his steam engine in 1849. Corliss was born in upstate New York, the son of a country doctor. By the standards of the day, he received a good education, but one that at the time, would not have included mechanical engineering. After graduating, he set up a general store.

STATIONARY STEAM ENGINE

Papin engine	1690
Savery engine	1698
Newcomen atmospheric engine	1712
Watt engine	1765
Watt double-acting engine	1784
Trevithick high-pressure engine	c. 1800
Corliss engine	1849

After three years, Corliss decided on a change of career that would use his interest in engineering. In 1842, he patented a heavy-duty sewing machine for shoes and leather goods. Two years later he moved to Providence, RI, where he hoped to find backing for his sewing machine. Getting a job as a draftsman, however, he soon found a new engineering project: the improvement of stationary steam engines, which, despite six decades of development, were still inefficient and expensive to run, and were generally used to power pumps for water mills.

"[By 1876] Corliss's design was recognized as one of the most significant of all American contributions to the development of steam engines."
HENRY'S ATTIC (2006) BY F. R. BRYAN AND S. EVANS

The centennial engine

Corliss started in business manufacturing with his improved stationary steam engine in 1848, patenting his revolutionary valve gear a year later. Corliss built his engines with standardized parts, which reduced their initial price and maintenance costs and made them affordable for mill and factory owners. But their major selling point was their economy as they were about 30 percent more fuel efficient than their rivals. For the first time, mills were no longer dependent on waterpower, and could move away from the millponds, canals and rivers on which they had once depended. Combined with a large immigrant workforce that was willing to work for low wages, Corliss engines were the foundation of U.S. industrial might in the late nineteenth and twentieth centuries.

The peak of Corliss's career was the selection of one of his engines to power the whole of the 1876 Centennial Exposition, held in Philadelphia, PA. The largest engine of its kind built during the nineteenth century, it stood 45 ft (14 m) high, with twin 44-in (1.1-m) cylinders that drove a flywheel 30 ft (9.1 m) in diameter to produce 1,400 hp. Corliss engines were so efficient, reliable and economical that several remain in use in the distillery industry in the twenty-first century.

A single giant Corliss steam engine powered the whole of the Centennial Exhibition, held in Philadelphia in 1876.

THE CORLISS STEAM ENGINE

At first sight, Corliss engines were stationary steam engines of standard design, with one or more pistons driving a flywheel that rotated at about 100 revolutions per minute. They were made in different sizes, the largest being the gigantic Centennial Engine, which stood at 45 ft (14 m) tall, with a flywheel 30 ft (9 m) across. They were used to power mills, finally ending the dependence on waterpower, and later to generate electricity. However, what made Corliss engines so superior to their predecessors and rivals was the Corliss valve gear.

KEY FEATURE:
CORLISS VALVE GEAR

In Corliss engines each cylinder is fitted with four valves, with inlet and exhaust valves located at each end. The cycle begins with the piston at one end of the cylinder, and the left exhaust valve and the right inlet valve both open. When the steam enters the cylinder, it pushes the piston toward the opposite end. Partway through the stroke, the right inlet valve closes. When the piston completes the stroke, the right exhaust valve and left inlet valve open, allowing the steam to the right of the piston to escape and steam to enter on the left side, pushing the piston. Partway through the stroke, the inlet valve closes, and the steam expansion pushes the piston to the end of the cylinder, completing the cycle.

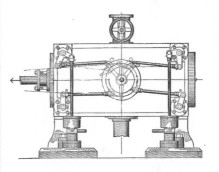

[A] Steam cylinder
[B] Wrist plate (valve gear)
[C] Steam inlet valve
[D] Steam exhaust valve
[E] Governor
[F] Crankshaft (valve gear eccentric)

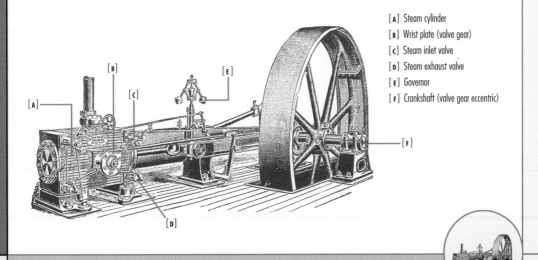

07

Designer:
Charles Babbage

BABBAGE DIFFERENCE ENGINE

Manufacturer:
Per Georg Scheutz

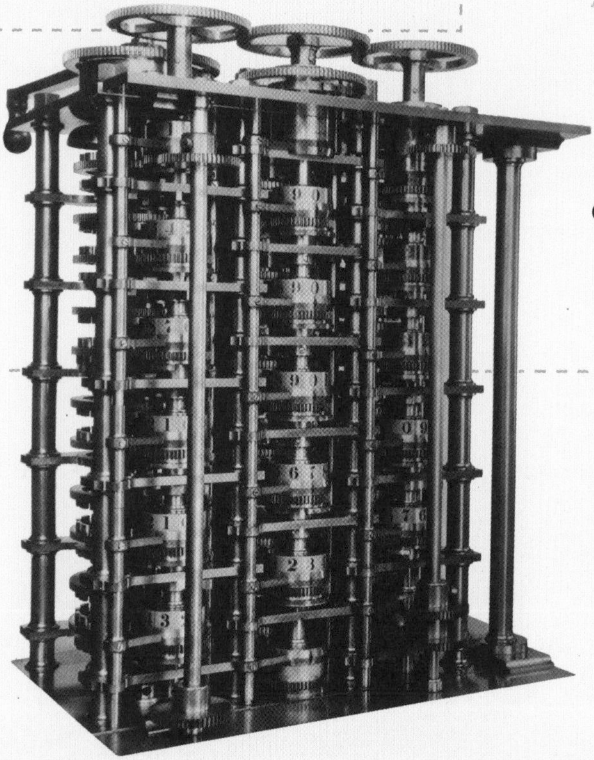

Industry
Agriculture
Media
Transport
Science
Computing
Energy
Home

1855

In the wake of the Industrial Revolution, scientists, engineers, financiers, surveyors and navigators needed accurate mathematical and astronomical tables to help them in their calculations. However, such tables were typeset by hand and full of errors, leading the mathematician and inventor Charles Babbage to devise several "engines" that would calculate and print the tables perfectly.

The steam-powered computer

This entry is unusual in that, of all the iconic machines featured in this book, it is the only one that was never built as conceived by its inventor, Charles Babbage (1791–1871). In fact, a completely faithful realization of his Difference Engine No. 2 was not achieved until the twentieth century, when London's Science Museum commissioned a full working machine for the bicentennial celebrations of Babbage's birth in 1991. The failure to build the engine was not for lack of funds, effort or enthusiastic support from his peers, however. The British government, not normally known for its liberality, sank the then considerable sum of £17,000 into the project, finally pulling the plug after 10 years when it became clear that the original machine would never be completed, as its inventor had moved on to design the even more ambitious Analytical Engine, which, had it been built, would have been the world's first programmable computer, pre-dating the 1936 German "Z1" by a century.

The interest of the British government, however, was not driven by philanthropy or the pursuit of pure science. At the beginning of the nineteenth century, engineers, accountants, bankers, surveyors, scientists, mariners and the military all had a pressing need for ready-made, accurate mathematical and astronomical tables to help them with their calculations. In his 1823 article on the Difference Engine, astronomer Francis Baily listed 12 mathematical tables, including logarithms, squares and cubes, and several astronomical tables used as navigational aids, which, though already in existence, he had found to contain multiple errors, mostly added at the typesetting stage. Babbage, a Cambridge professor of mathematics and prolific inventor, echoed Baily's concerns, when in 1821, he wrote in exasperation:

"I wish to God these calculations had been executed by steam."

Vive la difference!

The two Difference Engines (unglamorously numbered "one" and "two") were to be automatic, crank-operated mechanical calculators designed to create error-free tables by using Isaac Newton's (1642–1727) method of divided differences, which would then produce molds for printing plates. Babbage chose the method of divided differences because it obviates the need for multiplication and division, which are more difficult to achieve mechanically, and depends entirely on addition (subtraction being the addition of a negative number as in a modern computer). Babbage's Difference Engine No. 2 could store eight numbers 31 digits long and could accurately tabulate seventh-degree polynomials.

The method of divided differences relies on the fact that differentiation reduces the order of a polynomial by 1. If this is repeated then a zero-order polynomial can be obtained, i.e., a constant. The table below illustrates the method using the example of a second-order (or quadratic) polynomial. The first column shows consecutive values of x = 0, 1, 2, 3, 4; the second shows the corresponding values of p(x); the third contains the first difference d1(x) of the two neighbors in the second column; and the fourth, the second difference d2(x) of the two neighboring first differences. For a quadratic polynomial, the second difference is always constant, in this case 6. The table is easily constructed from left to right but, once the first few values are filled in, subsequent values can be obtained diagonally from top right to bottom left. To calculate p(5), start with the fourth column value of 6 (the constant), add it to 19 in column three to get 25, then continue to the second column and add 42. Thus p(5) is 25 + 42 = 67. Repeating these steps gives p(6) = 98, and so on.

x	$p(x) = 3x^2 - 2x + 2$	$d1(x) = p(x+1) - p(x)$	$d2(x) = d1(x+1) - d1(x)$
0	2	1	6
1	3	7	6
2	10	13	6
3	23	19	
4	42		

Tabulated values for $p(x) = 3x^2 - 2x + 2$ (courtesy of Dr. D. Scott)

Building Babbage's engines

Before the Science Museum's successful realization of the Difference Engine No. 2, there were several attempts to build difference engines, starting with Babbage's own. In 1823, having obtained £1,700 from the British government, Babbage engaged Joseph Clement (1779–1844), one of the finest draftsmen and toolmakers of his generation to build the Difference Engine No. 1. Socially and temperamentally the two men could not have been more different: the touchy, academic Babbage and the blunt, unpolished Clement. Although the two men fell out over money in 1832, and the government withdrew its funding bringing an end to the project, Clement and Babbage managed to produce a demonstration piece that is one of the finest examples of high-precision engineering from the early nineteenth century. During the 1850s two enterprising Swedes, Per Georg Scheutz (1785–1873) and his son Edvard, built several versions of the Difference Engine.

Scheutz's version of the Difference Engine was far inferior to Babbage's original concept.

The Scheutz versions were much smaller than Babbage's, fitting on a tabletop rather than requiring a whole room, but technically and mathematically they were far inferior. The best Scheutz engine could only store four numbers of 15 digits. One of the machines was sold to the London General Register Office in 1859, but it lacked many of the security features envisaged by Babbage, was difficult to work, and often broke down. Instead of making their fortunes with the Difference Engine, the Scheutzes went bankrupt.

COMPUTING

Schickard calculator	1623
Pascaline	1642
Jacquard loom	1801
Babbage Difference Engine No. 1	1832
Babbage Analytical Engine	1834
Babbage Difference Engine No. 2	1847
Scheutz Difference Engine	1855

1791–1871

Charles Babbage

BABBAGE DIFFERENCE ENGINE

ANATOMY OF . . .

THE BABBAGE DIFFERENCE ENGINE (NO. 2)

[A] Columns with number wheels, sector gears, and carry levers
[B] Crank mechanism
[C] Crankshaft

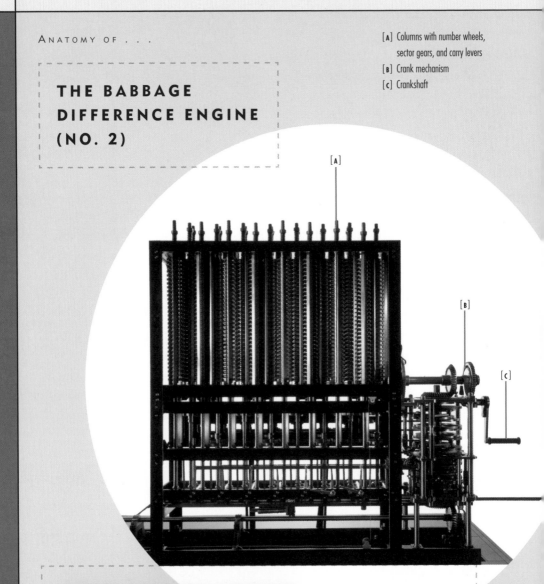

[A]

[B]

[C]

Joseph Clement fashioned each component of the Difference Engine from brass—a process that demanded a high degree of precision.

Babbage's engine was the finest example of high-precision engineering in the early nineteenth century.

The Difference Engine has three main sections: the columns carrying the number wheels (0–9, divided into odd and even numbers), sector gears, and carry levers; the hand crank; and the "printer." The columns are labeled 1 to n, and each can store one number. Column n always stores a constant (see table on page 32), while column 1 displays the value of the calculation. After setting the initial values to the columns, the rest of the operation becomes automatic. In order to complete one full set of additions, the crank has to be turned four times, executing the following four steps: (Step 1) All even columns are added to the odd columns, and the even columns are returned to zero. As they turn to zero, the wheels transfer their values to sector gears between the columns, which are then added to the odd columns. If an odd column reaches zero, a carry lever is activated; (Step 2) as carry propagation is accomplished by a set of spiral arms, the sector gears return to their original positions, causing them to restore the values to the even columns; (Step 3) this mirrors Step 1, but the odd columns are added to the evens. The sector gears transfers the column 1 values to the printer; (Step 4) like Step 2, this returns the odd columns to their original values.

KEY FEATURE:
THE PRINTER

Although called a "printer," the primary purpose of this part of the engine was to produce stereotype plates for use in printing without the need for hand compositing that introduced errors into the tables. The device had some very sophisticated features, including variable line height, number of columns and column margins, as well as automatic line or column wrapping. The ink-and-paper printout was used to check the engine's output.

08

Designer:
Isaac Merritt Singer

SINGER "TURTLE BACK" SEWING MACHINE

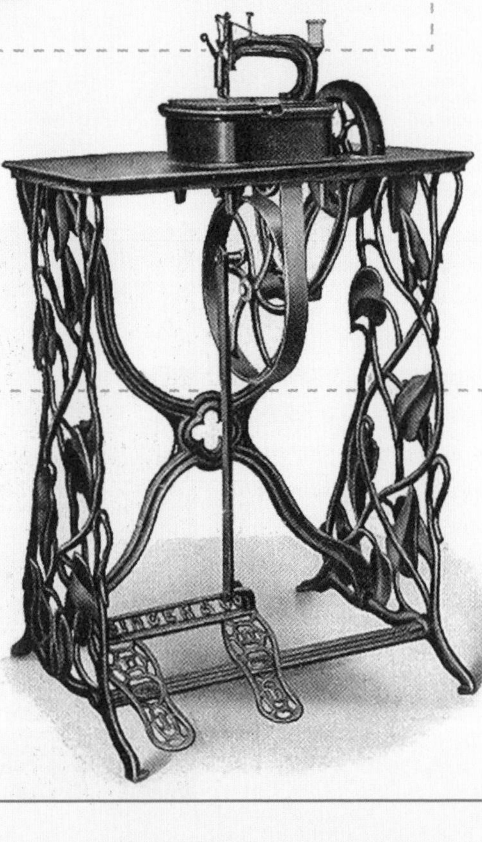

Industry
Agriculture
Media
Transport
Science
Computing
Energy
Home

Manufacturer:
I. M. Singer and Co.

1856

Isaac Merritt Singer did not invent the sewing machine but he was the industry's best early marketeer. With the "Turtle Back," the first sewing machine manufactured specifically for the domestic market, he established Singer as the industry leader and created the U.S.'s first multinational corporation.

Isaac Merritt Singer

Stitched up

With this entry on the "Turtle Back" sewing machine, we encounter one of the most colorful characters featured in this book. Inventors are often "eccentrics," but their eccentricities are, for the most part, limited to their fields of endeavor. In Isaac Merritt Singer (1811–1875), we find a man who was truly larger than life. For one thing, by Victorian standards, he was a giant: standing 6'4" (1.95 m), with appetites that matched his large frame. Although trained in his brother's machine shop, Singer had very different ambitions: to become an actor. In 1839, having sold his first engineering patent, he set up his own touring company, the Merritt Players, who stayed together until the money ran out. Singer was also a serial womanizer, with five "families" and a bigamous marriage that forced him to leave the U.S. and settle in England.

"You want to do away with the only thing that keeps women quiet—their sewing!"

ISAAC SINGER COMMENTING TO HIS PARTNER ON HIS OWN INVENTION.

Hunting around for another invention to make his fortune, Singer took out his first sewing machine patent in 1851. However, it was Elias Howe (1819–1867) who held the American patent for the sewing machine. He sued Singer and other manufacturers for patent infringements, triggering the "Sewing Machine Wars," which were finally settled by the 1856 "Sewing Machine Combination," consisting of Howe, Singer, and two other manufacturers, who agreed to pool their patents rather than engage in further expensive litigation. That same year, Singer marketed the Turtle Back, the first machine aimed at the domestic market, and the first to be powered by a treadle that left the operator's hands free. Although the machine initially cost $100, which put it out of reach of most American pocketbooks, Singer devised an installment plan so that a customer could take her machine home for a down payment of $5. Although the Turtle Back was soon replaced, it established Singer as the industry leader. By the 1870s, he had opened factories all over the world and created the U.S.'s first multinational corporation.

SINGER "TURTLE BACK" SEWING MACHINE

09

Designer:

Isambard Kingdom Brunel

SS *GREAT EASTERN*

Manufacturer:

J. Scott Russell & Co.

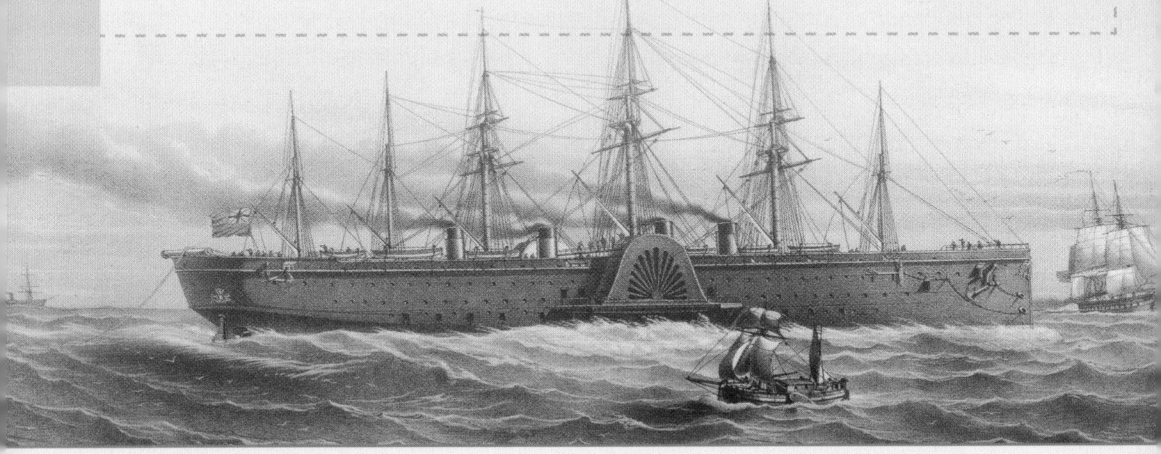

Industry

Agriculture

Media

Transport

Science

Computing

Energy

Home

"No ship even vaguely approaching the *Great Eastern* in size or complexity had been built before, yet she was commissioned to be built in a traditional shipyard, albeit by a shipbuilder of outstanding reputation in the shape of John Scott Russell."

R. Buchanan, *Brunel* (2006)

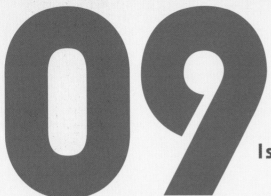

The bottom of the ocean is littered with the wrecks of "unsinkable" ships. Although the SS *Great Eastern* did not sink, Brunel's largest steamship, which was meant to be the crowning achievement of his career, was dogged with problems and mishaps almost from the day of her conception. Nevertheless, with a double hull and powered steering gear, she set the standards for the construction of later liners, and her great size made her suitable for the task of laying the first successful transatlantic telegraph cable.

Isambard Kingdom Brunel

The "Great Babe"

One only need say the name RMS *Titanic*, which sank during her maiden voyage in 1912 (see "Marconi radio," page 72), to conjure up images of ill-fated giant passenger liners—vaunted by their designers to be the largest yet built and "unsinkable"—until, that is, they sank. Perhaps Poseidon, the God of the Sea, is particularly touchy about this form of human hubris. Fortunately the *Great Eastern*, although she was dogged with problems from the day of her conception and throughout her short career as a passenger liner, did not sink on her maiden voyage, although a boiler explosion killed five stokers and injured many others on her first outing in 1859, and at the end of her 1862 transatlantic crossing, she was holed by rocks off Long Island, opening a gash in her outer hull that was 60 times greater than the one that caused the *Titanic* to sink. Unlike the tragic *Titanic*, the *Great Eastern*, which her designer, Isambard Kingdom Brunel (1806–1859), nicknamed his "Great Babe," remained afloat thanks to her double hull, and made it to New York for repairs under her own steam.

The *Great Eastern* was the most ambitious steamship project of the nineteenth century, holding the record for the largest ship afloat for four decades, and for the greatest tonnage until the early twentieth century. Models of her show a strange hybrid of different maritime technologies: She had six masts for sails, and funnels for her steam engines that powered two giant paddle wheels and a screw propeller. She had holds for freight and luxurious passenger accommodation, having been designed for the long passage to India and Australia, which she could reach without the need to refuel. However, after the bankruptcy of her builder and other mishaps, the great ship was switched to sail on the transatlantic route. In 1865, having failed to make money as a liner, she was converted to lay transoceanic telegraph cables, a task that she performed until 1878. After a further 10 years as a showboat and tourist attraction in England, she was broken up for scrap.

The man in the stovepipe hat

Unlike that earlier pioneer of the steam age, George Stephenson (1781–1848), Brunel benefited from an excellent education, first in England and then in France (thanks to his French father), followed by an apprenticeship with one of France's top clockmakers. Like Stephenson, however, he is best remembered for his railroads. He built the engineering marvel of the age, the Great Western Railway (GWR), from London to Bristol, which included some of the country's most innovative tunnels and bridges, including the Clifton Suspension Bridge. Brunel, always pictured in his trademark stovepipe hat, did not see why the passenger service should stop at the port of Bristol, the terminus of the GWR. He envisaged a service that would take passengers from Paddington Station in London to Bristol, and then on by steamship to the booming New World, which was welcoming tens of thousands of European immigrants.

The *Great Eastern* was the third and most ambitious steamship designed and built by Brunel. In 1838, he launched the 252-ft (77-m) SS *Great Western*, a wooden-hulled paddle steamer, designed for the transatlantic route. His second ship was the 322-ft (98-m) SS *Great Britain*, which was the first to combine an iron hull and a screw propeller, which completed her maiden voyage in 1843. The *Great Eastern* would be even larger, and although not an unqualified success as a liner, she set the standards for the construction of passenger liners for decades to come.

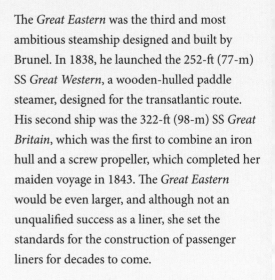

STEAMSHIPS

Pyroscaphe	1783
Comet	1812
Savannah	1819
James Watt	1820
Great Western	1837
Great Britain	1843
Great Eastern	1858

Brunel's ambitions for the *Great Eastern* were never matched by her performance.

ANATOMY OF . . .

THE SS *GREAT EASTERN*

At 692 ft (211 m), the *Great Eastern* was more than twice the length of the *Great Britain*, and 190 ft (58 m) shy of the *Titanic*'s 882 ft (269 m), with a beam of 82 ft (25 m) and a displacement of 32,160 tons. She was powered by five steam engines: four for the 56-ft (17-m) paddle wheels, and one for the 24-ft (7.3-m) screw propeller, that combined gave her a top speed of 14 knots (16 mph; 26 km/h). In addition, she had six masts rigged for sails, but in practice, these could not be used with the engines, because the hot exhaust from the funnels would have set fire to the canvas. She could carry 4,000 passengers (compared to the *Titanic*'s 2,453), and had a crew complement of 418. She was able to carry enough coal for a round trip from England to Australia, and was equipped with large cargo holds for freight. Technically she had two main innovations, a powered steering engine, as it was impossible to steer a ship of her size by hand, and a double hull, which saved her from sinking when she was holed off the American coast.

KEY FEATURE: THE DOUBLE HULL

The Great Eastern was built with a revolutionary double hull, one inside the other with a crawl space of 2 ft 10 in (86 cm), and strongly braced every 6 ft (180 cm). The hulls were made of standardized iron plates ¾ in (19 mm) thick that were riveted together with standard-sized rivets—the first time standardization had been applied to such a huge engineering project.

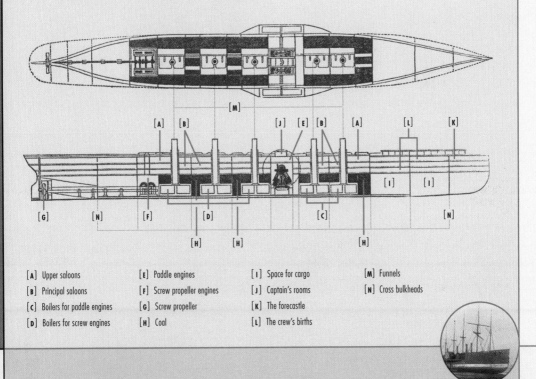

[A] Upper saloons
[B] Principal saloons
[C] Boilers for paddle engines
[D] Boilers for screw engines
[E] Paddle engines
[F] Screw propeller engines
[G] Screw propeller
[H] Coal
[I] Space for cargo
[J] Captain's rooms
[K] The forecastle
[L] The crew's births
[M] Funnels
[N] Cross bulkheads

SS GREAT EASTERN

10

Designer:

John Wesley Hyatt

HYATT STUFFING MACHINE

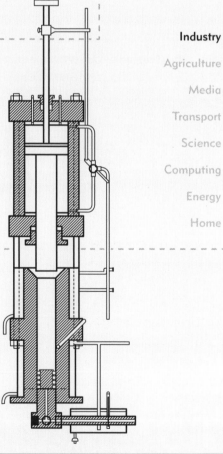

Manufacturer:

Celluloid Manufacturing Company

Industry ■

Agriculture

Media

Transport

Science

Computing

Energy

Home

"[Celluloid's] applications included everyday items like combs, dentures, knife handles, toys and spectacle frames."

P. PAINTER AND M. COLEMAN, *ESSENTIALS OF POLYMER SCIENCE AND ENGINEERING* (2009)

1872

When the Hyatt brothers patented their unfortunately named "Stuffing Machine" and "Celluloid," they established the plastics industry that would transform the world in the twentieth century.

How Celluloid saved the elephant

By the middle of the nineteenth century, such was the demand for ivory—for billiard balls, as well as smaller items such as buttons, knife handles, piano keys, false teeth, fan ribs and collar stays—that even the vast African elephant herds were under threat of extinction. With the world ivory shortage affecting their business, Phelan & Collander, an American manufacturer of billiard balls, offered a $10,000 prize to anyone who could produce an ivory substitute. The inventor John Wesley Hyatt (1837–1920) began to experiment with "Parkesine," the first man-made plastic, invented by British inventor Alexander Parkes (1813–1890) in 1862. During the 1860s, Parkes had tried and failed to commercialize his invention, but one of his associates, Daniel Spill (1832–1887), had more success with an improved version of the plastic that he named "Xylonite" in 1869. In 1870, John Hyatt and his younger brother Isaiah had come up with their own version of Parkesine, which they christened "Celluloid." As was often the case with nineteenth-century inventions, the rival claims led to protracted legal action over patent infringements that ultimately recognized Parkes' claims as the original inventor.

Although now forever associated with the movie industry, celluloid, a compound of nitrocellulose and camphor, was the world's first commercially successful plastic. Instead of claiming the $10,000 prize, the Hyatt brothers set up their own company to manufacture billiard balls and other products formerly made of natural materials such as horn and ivory. The real secret of the brothers' success was not just the development of celluloid but also their invention of the first injection molding machine—the "Stuffing Machine"—that could produce bars or sheets of celluloid that were then worked into finished products. Patented in 1872, the Stuffing Machine was based on a metal pressure die-casting machine. The Hyatts added a mandrel in the barrel to improve the heat conductivity to melt the celluloid, which was then injected into a water-cooled mold by a plunger mechanism.

HYATT STUFFING MACHINE

11

Designer:
Zénobe Gramme

GRAMME MACHINE

Manufacturer:
Zénobe Gramme

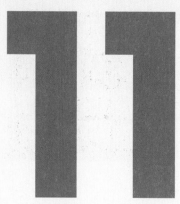

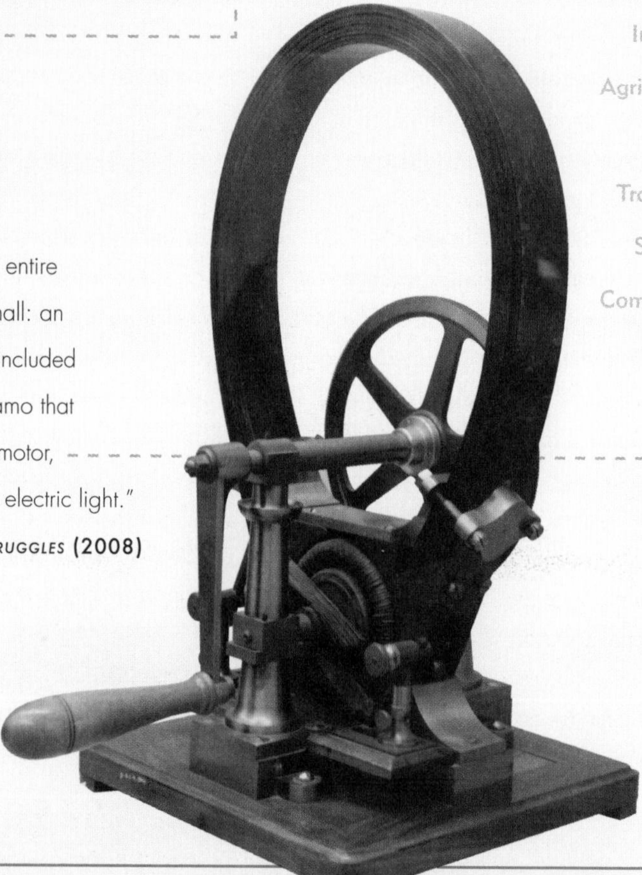

Industry
Agriculture
Media
Transport
Science
Computing
Energy
Home

"Gramme presented an entire electrical world writ small: an integrated system that included a steam-powered dynamo that furnished current for a motor, electroplating, and the electric light."

M. SCHIFFER, *POWER STRUGGLES* (2008)

1873

The Gramme machine, the first commercially viable electric motor working on direct current (DC), began the transformation of the First Industrial Revolution's world of mechanical steam power and gas lighting into the world of the Second Industrial Revolution, powered and lit by electricity.

Electrifying the world

The ancients knew of "electricity," although it only acquired this name in 1600 (from the Greek *elektron*, meaning "amber," because static electricity could be produced by rubbing amber). However, the phenomenon remained little understood until the discoveries made by several nineteenth-century scientists, notably Hans Ørsted (1777–1851) and Michael Faraday (1791–1867). With this new understanding came attempts to commercialize electricity as a source of power and light for industry and the home. At first, steam-powered magnetos and dynamos were used to generate electricity, but due to poor design, these produced low, intermittent outputs. In the early 1870s, the Belgian inventor and former cabinetmaker Zénobe Gramme (1826–1901) designed and manufactured particularly efficient dynamos. He produced two steam-powered models: a low-voltage one-horsepower machine designed for electroplating and a high-voltage four-horsepower dynamo for lighting that outshone its larger but less efficient rivals.

In 1873, when Gramme was displaying his dynamos at an industrial exhibition in Vienna, his assistant accidentally connected the output wires of two dynamos. When a steam engine began to turn the first dynamo, it caused the armature of the second to rotate at speed. Gramme realized that the second dynamo had become a powerful electric motor, outperforming all existing devices. Gramme immediately capitalized on this happy accident by rigging up a new attraction for the exhibition: He connected two dynamos almost 1 mile (1.6 km) apart, with the "motor" pumping water for a small waterfall. With this simple display, Gramme demonstrated two principles that would go on to transform the world: First, with a few modifications, his dynamos could be converted into electric motors that could drive industrial machinery; and second, and even more significant, he had shown that mechanical power could be generated in one location, converted into electricity with a dynamo, transmitted over long distances by wires and turned back into mechanical energy by electric motors at the other end.

12

LINOTYPE TYPESETTING MACHINE

Industry

Agriculture

Media ■

Transport

Science

Computing

Energy

Home

Manufacturer:

Mergenthaler Linotype
Company

1884

After the invention of movable type in fifteenth-century Europe, the world had to wait four centuries for the next revolution in typesetting. The Mergenthaler Linotype machine speeded up the production of books and newspapers, slashing production costs and making printed materials available to a much larger readership than ever before.

Making an impression

Today we have become used to typing text straight onto a computer screen, adjusting font, size, bold, italics, spacing and justification with a few keystrokes, and sending the document to a printer from which—at least in theory—it emerges perfectly printed. Six centuries ago European printers laboriously carved whole pages onto blocks of wood, and it was only in the fifteenth century that the German printer Johannes Gutenberg (c. 1398–1468) created the first movable type in Europe (as usual, the Chinese had got there centuries earlier).

Gutenberg's system allowed pages of text to be assembled from individual and interchangeable characters carved onto small pieces of lead. Despite being a major step forward from woodblocks, the process of typesetting books, magazines and newspapers remained time-consuming, requiring the labor of dozens of skilled operators called "compositors." Books, though more widely available then ever before, remained comparatively expensive and beyond

the budget of large sections of society, and newspapers were limited in length to eight pages. In short, the high costs and limitations of typesetting technology continued to stifle the dissemination of knowledge and information.

For the many autocratic regimes that held power in the nineteenth century, this was actually a desirable state of affairs, because an ignorant citizen was much easier to control than an educated and well-informed one. However, the Second Industrial Revolution that began in the closing decades of the nineteenth century transformed the world with the invention of new media and communication technologies, notably the wireless telegraph, the typewriter and the telephone. Printed media, however, remained the main channel through which people would be educated and informed, and would be so for decades to come. The time was ripe for an advance in the process of compositing, or typesetting, which had not changed significantly since the days of Gutenberg.

LINOTYPE TYPESETTING MACHINE

The Second Gutenberg

The man who revolutionized typesetting in the nineteenth century and earned the nickname "the second Gutenberg" was not like many of today's innovators—grad students from MIT or Cambridge University, or researchers employed by large multinational corporations. He was an inspired amateur inventor who saw a good idea and found practical ways of realizing it. Ottmar Mergenthaler (1854–1899) was the third son of a German schoolteacher, who was apprenticed to a watchmaker in his native Germany. At the age of 18, he emigrated to America, like so many other economic migrants before and after him, to escape the limited prospects in his homeland and realize his own "American dream."

In 1876, James O. Clephane (1842–1910) and Charles T. Moore (1847–1910) approached the 22-year-old Mergenthaler, then a partner in a scientific instrument business in Baltimore, MD, with a design for a typesetting machine. Clephane and Moore were looking for a way to speed up the publication of court reports and had based their design on the typewriter, an invention that had been successfully commercialized in the U.S. a few years earlier (see page 84). Their machine, however, suffered from many design flaws, including its use of papier-mâché molds, which did not produce clear type impressions.

They asked Mergenthaler if he could make their design work. It took 8 years and several complete redesigns of the Clephane-Moore prototype before Mergenthaler produced the first machine combining a modified typewriter keyboard with a hot-metal casting function that produced an entire line of text. He patented his invention in 1884, and it had its first commercial demonstration in July 1886, in front of Whitelaw Reid (1837–1912) the editor of the *New York Tribune*, who exclaimed, "Ottmar, you've done it again! A line o' type!"—from which the machine got its name: the Linotype.

"The marvel of the century, combining the casting, setting and distributing of type in one machine, at one operation, by one operator."

PROMOTIONAL MATERIAL FROM THE MERGENTHALER LINOTYPE CO., (1895)

Ottomar Mergenthaler 1854–1899

By automating a process that had required skilled compositors, the Linotype dramatically reduced the cost of printed matter.

Schoolbooks and tabloids

Mergenthaler's machine looked a bit like a giant typewriter, standing 7 ft (2 m) tall by 6 ft (1.8 m) wide, but it produced much more than text printed in ink on paper. Today's mass-media culture is largely the product of the Linotype revolution of the late nineteenth century. Mergenthaler himself had once complained of the lack of schoolbooks in his native Germany when he was a child, and one of the greatest boons of his invention was to make schoolbooks much more readily available, hence transforming and standardizing education all over the world. Of course, the availability of cheaper books was not limited to education: it also made works of fiction and non-fiction available to the general public. This included both the great works of world literature and penny-dreadfuls and pulp fiction novels; and books on science, engineering, politics and economics that stimulated the technological and ideological revolutions of the twentieth century. One of the unintended consequences of the Linotype was the growth of the tabloid press on both sides of the Atlantic. One wonders what Mergenthaler would have made of the dubious use to which his invention would be put when it was used to typeset the stories found in tabloids such as the UK's *The Sun* and the U.S.'s *National Enquirer*.

PRINTING AND TYPESETTING

Woodblock printing	200
Movable type (in China)	1040
Movable type (in Europe)	1454
Etching	c. 1500
Lithography	1796
Rotary press	1843
Offset printing	1875
Linotype	1884
Phototypesetting	1960s
Digital press	1993

LINOTYPE TYPESETTING MACHINE

THE LINOTYPE

[A] Keyboard
[B] Magazine
[C] Assembler
[D] Casting mechanism
[E] Distributor

[E]
[B]
[C]
[D]
[A]

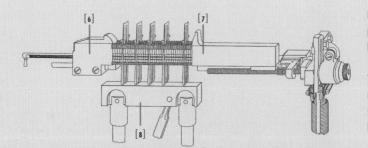

[6]
[7]
[8]

Diagram of the justification process. The composed line is locked up between the jaws — [6] and [7] — of the vise. The justification ram [8] then moves up to expand the spacebands to fill the space between the vise jaws.

The Linotype consists of five major sections: keyboard, magazine, assembler, casting mechanism, and distributor. The keyboard has 90 keys divided into three sections: lower-case letters are black; upper-case letters are white; and numbers, punctuation and other characters are blue. The characters are arranged in order of frequency of use and not in the modern "qwerty" arrangement of a computer keyboard. There is also a spaceband lever—the equivalent of the spacebar on a modern keyboard.

Up to four magazines sit on top of the keyboard and these contain the letter matrices. Each matrix stores the letterform for a single character in a particular type design and size, in both roman and italic. Each magazine holds one font, and the operator can switch between magazines to mix fonts in a single line of text. When the operator hits a character on the keyboard, the appropriate matrix travels from the magazine, through a channel, into the assembler. Once a whole line is completed, the operator pulls a lever to send the line into the casting mechanism. The line is cast in an alloy of lead, tin and antimony, and produces a one-piece line casting, or "slug," that can be impressed up to 300,000 times before it begins to develop imperfections and has to be re-cast.

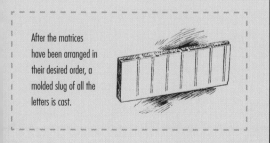

After the matrices have been arranged in their desired order, a molded slug of all the letters is cast.

Right: Linotype matrices stacked together.

KEY FEATURE:
DISTRIBUTION MECHANISM

One of the Linotype machine's greatest labor-saving features is the distribution mechanism that returns the matrices and spacebands to their respective storage areas. This is accomplished by carving a pattern of teeth on top of the matrices to identify each character, so that they can be correctly sorted, and made available to the operator for the next line of type.

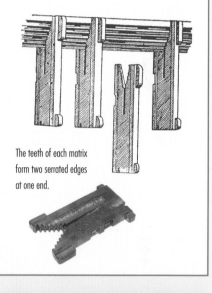

The teeth of each matrix form two serrated edges at one end.

13

PARSONS' STEAM TURBINE

Manufacturer:

C.A. Parsons and Company

Industry

Agriculture

Media

Transport

Science

Computing

Energy

Home

1884

With a daring naval stunt, Charles Parsons demonstrated that his steam turbine could beat any ship fitted with a conventional steam engine. Yet the real significance of the steam turbine was not faster ships but a vastly increased and cheaper electricity supply.

"Gatecrashing" Queen Victoria's Diamond Jubilee

In 1897, Queen Victoria (1819–1901) celebrated 60 years on the throne with a review of the British imperial fleet, then the strongest and largest in the world, at Spithead off the English south coast. In full view of the Queen, the Prince of Wales and assembled government and naval dignitaries, a small launch, the *Turbinia*, sped into view among the great ships of the fleet. She easily evaded interception by other vessels because she literally left them standing in the water. Today, with modern concerns about terrorism, a similar "gatecrasher" would be blown out of the water, but in that less troubled time, the stunt allowed the builder of the *Turbinia*, Charles Parsons (1854–1931), to demonstrate the superiority of his marine steam turbine over conventional reciprocating steam engines.

"It seemed to me that moderate surface velocities and speeds of rotation were essential if the turbine motor was to receive general acceptance as a prime mover." CHARLES PARSONS IN A LECTURE IN 1911

The steam turbine was not a new idea but it had long been considered impracticable. No less a figure than the father of steam power, James Watt (1736–1819), thought that a steam turbine could not be built because of the huge centrifugal forces caused by steam traveling through an engine at 1,700 mph (2,735 km/h). Parsons' solution, patented in 1884, was to build a staged design that reduced and controlled the speed of the steam as it passed through. He also employed the reactive power of the steam as it left one section to drive the turbine blades. His turbine was so much more efficient than a conventional piston-driven steam engine that it transformed electricity generation. Hitherto, dynamos had been driven at between 1,000 and 1,500 rpm, but Parsons realized that his turbine could drive a dynamo at the then unheard of velocity of 18,000 rpm. Parsons built his own turbo-generators, which he began to install all over the British Isles and then worldwide. In 1923, he was awarded the contract for the world's largest electricity plant, producing 50,000 Kw for the city of Chicago.

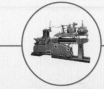

14

"ROVER" SAFETY BICYCLE

Manufacturer:

Starley & Sutton Co.

Industry

Agriculture

Media

Transport ■

Science

Computing

Energy

Home

1885

Although the bicycle had existed in some form since the early nineteenth century, it was not until the late nineteenth century that it began to transform society. In addition to providing inexpensive transport for commuting and leisure, the Rover safety bicycle played an important role in the emancipation of women during the early years of feminism.

The wooden horse

The modern history of the bicycle begins just after the Napoleonic Wars (1803–1815) with the "velocipede"—a wooden frame fitted with two wheels powered by the rider pushing with his feet along the ground. Difficult to steer and a bumpy ride on the unpaved roads of the day on its twin metal-rimmed wheels, the velocipede was originally conceived as a replacement for the horse. The cavalries of Europe kept their horses, and it was the fashionable young men of London and Paris who took to the machine, giving it its other name: the "dandy-horse." For the next 68 years, the bicycle underwent several strange transformations: gaining extra wheels, with the development of tricycles and quadra-cycles, and using various forms of power transmission, including pedals on the front wheel crank and treadles.

"Safer than any Tricycle, faster and easier than any Bicycle ever made. Fitted with handles to turn for convenience in storing or shipping. Far and away the best hill-climber in the market."

1885 AD COPY FOR THE ROVER SAFETY BICYCLE

From the modern point of view, the oddest was what was called the "ordinary bicycle" during the 1870s and 1880s, but is better known as the "penny-farthing" from its oversized front wheel. Dangerous to ride and difficult to steer, the penny-farthing was popular because its large wheel made it faster and more comfortable on unpaved roads than the velocipede; but at the same time it made cycling unsuitable for anyone other than the daring and physically fit. When the British bicycle manufacturer John Kemp Starley (1854–1901) marketed the "Rover" in 1885, he was going back to earlier cycle designs, such as the velocipede and the "bone-shaker." The rider was once again within reach of the ground, which made balancing, steering, mounting and dismounting much simpler and safer, hence the epithet given to the Rover: the "safety bicycle." Although the Rover was initially more expensive, less comfortable and heavier than the penny-farthing, it was an invention that had found its time. Within a decade, and after the invention of inflatable tires had made it a smoother ride, it quickly became the world's standard bicycle.

"ROVER" SAFETY BICYCLE

"Like a fish needs a bicycle"

The invention of the safety bicycle casts a new light on the feminist quip about a woman needing a man "like a fish needs a bicycle." Ironically, the safety bicycle played a major role in the emancipation of women and the creation of the late-nineteenth century's "new woman." Although women had ridden tricycles and quadracycles, the modest Victorian dress code made it impossible for them to ride a penny-farthing. However, even in a floor-length skirt, women were able to ride a safety bicycle. Thus, the long fight for women's political and social emancipation began when they took to the streets on bicycles, giving them unprecedented mobility, self-reliance and independence.

One of the most extraordinary representatives of this new breed of women was Annie Kopchovsky (1870–1947), who, under the name Annie Londonderry (for her commercial sponsor), was the first woman to cycle around the world in 1895, completing the trip in 15 months. What was more shocking to the Victorians than a woman cycling around the globe was that she did most of the trip dressed in voluminous cycling "bloomers." Although about as revealing and sexy as a workman's overall, bloomers nevertheless showed that the wearer had legs, making Annie the Lady Gaga of her day.

The Rover was called a "safety bicycle" because the rider's feet were within reach of the ground.

BICYCLE

Velocipede	1817
Treadle velocipede	1839
Quadracycle	1853
Bone-shaker	1863
Penny-farthing	1869
Rover safety bicycle	1885

The 1880 "penny-farthing" and the 1886 "safety bicycle."

THE ROVER
SAFETY BICYCLE

[E]

[F]

[A]

[B]

[D]

[C]

[A] Steerable front wheel
[B] Diamond frame
[C] Chain and chain ring
[D] Rear wheel and sprocket
[E] Adjustable handlebars
[F] Adjustable seat

KEY FEATURE: THE WHEELS

It may seem extraordinarily obvious to us that two wheels of the same size, driven by a chain and pedal assembly, is the safest and most efficient configuration for a bicycle, but it took nearly half a century for manufacturers, led by Starley, to return to the velocipede's original equally sized wheels.

Although none of the individual components of the Rover was completely original, by assembling them in one design, Starley created the modern bicycle, whose general form and major features have changed little since 1885. The diamond frame, now a standard cycle feature, was found to be the strongest and most comfortable configuration once pneumatic tires had been introduced. The two wheels were nearly the same size (the original Rover's front wheel was slightly larger), and the forks were set at an angle, making the bike much easier to steer. The drive was a chain attached to the rear wheel and turned by foot pedals, replacing the pedals on the crank of the front wheel as in the penny-farthing, which made it difficult to steer and pedal at the same time.

15

WESTINGHOUSE AC SYSTEM

Manufacturer:

Westinghouse Electric

Industry

Agriculture

Media

Transport

Science

Computing

Energy ■

Home

"One of [AC's] major selling points was that [it] would allow long-distance transmission of power, whereas Edison's DC system would not." M. SCHIFFER, *POWER STRUGGLES* (2008)

1887

Nikola Tesla

During the 1890s, Westinghouse's AC and Edison's DC were locked in an epic battle for dominance of the world's electricity generation and supply industry. Despite Edison's genius for marketing and advertising, and the electrocution of an elephant, AC finally won the day.

Deep-fried Dumbo

One of the least edifying episodes in the "war of the currents" between AC (alternating current) and DC (direct current) was the execution of Topsy, a 28-year-old elephant from Coney Island that had trampled three men to death (including one who had fed her a lit cigarette and got exactly what he deserved). Although hanging had been considered, Thomas Edison (1847–1931), the major proponent and purveyor of direct current in the U.S., suggested that Topsy be electrocuted with his rival George Westinghouse's (1846–1914) alternating current. In doing so, he hoped to discredit AC, going as far as calling execution by electrocution being "Westinghoused." A decade earlier, Edison had scored an even greater publicity coup by ensuring that the electric chair, first used in New York in 1890, also used AC. On January 4, 1903, Topsy was fed 16 oz (460 g) of cyanide (just in case) and a 6,600-volt charge was passed through her body. As a film Edison made of the event (which is available online) shows, the elephant died almost instantaneously, and, one hopes, painlessly.

Although "Electrocuting an Elephant" was shown worldwide—the early twentieth-century equivalent of a video going viral—Edison and DC had already lost the war of the currents, as by the late 1890s, the world's major industrial nations were investing heavily in AC. Designed for Westinghouse Electric by the brilliant if slightly unstable Serbian engineer Nikola Tesla (1856–1943), the Westinghouse AC system was far more versatile, efficient and economic than its DC rival promoted by Edison. Tesla had once worked for Edison, but the older man had summarily dismissed Tesla's proposals for AC generation and transmission, it is said because Edison did not have the mathematical knowledge fully to understand its underlying principles. Hence, the rivalry was not just between two technologies but also between the principals involved on each side: Edison versus Westinghouse and Tesla, who enthusiastically lambasted each other in the press.

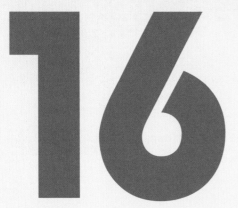

16

Designer:
Emile Berliner

BERLINER "GRAMOPHONE"

Manufacturer:
Berliner Gramophone Co.

Industry
Agriculture
Media
Transport
Science
Computing
Energy
Home

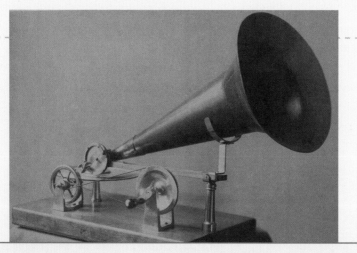

1887

Emile Berliner

Although the "phonograph" and cylinder predated the "gramophone" and record disc by over a decade, it was the gramophone that won the recording industry's first format war. The cylinder is now a historical curiosity, while for the enthusiast the vinyl disc still represents the acme of sound reproduction.

Nineteenth-century format wars

In recent decades, we have become used to sound-recording formats succeeding one another at a dizzying pace. I am of the generation that grew up with vinyl records (No, not 78s! 45s and LPs), and in my lifetime, I have seen reel-to-reel tape (page 150), cassette tape (see "Walkman," page 192), 8-track, CD, DAT and MiniDisc, each hailed as the ultimate in sound reproduction and now all made obsolete by online digital formats. The conventional gramophone disc, invented by Emile Berliner (1851–1929) in 1888 and first marketed a year later, lasted over seven decades, while tape and optical disc formats have managed a few decades each. However, at the very beginnings of the sound-recording industry, we find a similar succession of formats, until the recording industry settled on the disc as the common standard. The late nineteenth-century equivalent of the vinyl LP versus the CD was the cylinder versus the gramophone disc.

As every school child knows, the prolific inventor Thomas Edison (1847–1931) invented the "phonograph" in 1877. The real story, however, is a bit more complicated, with at least two previous inventors who had succeeded in devising methods to record sound but not to play it back. However, Edison, with his usual flair, was the first to patent the idea and bring a finished product to market. His phonograph, which could both record and play back, did not use discs but cylinders—initially made of fragile tin foil, which was later replaced by wax. The cylinder has several advantages in terms of constant playback speed, and it did not disappear until 1929, when Edison finally withdrew it, acknowledging the victory of the disc. Ten years after Edison's phonograph, Emile Berliner patented the "gramophone," which at first also used a cylinder but, shortly after, Berliner had produced his first recordings on disc.

"To return to the main menu, press one . . ."

The phonograph owes its invention to the telephone, because Edison was initially looking for a way to record and play back telephone messages. Had he realized his idea, we would have had the joy (or, rather, horror) of automated telephone services a century early. Fortunately for the world, recording technology was not sufficiently developed and humanity was spared telephone automation until the twentieth century. Berliner, too, was interested in the telephone. Like the inventor of the Linotype, he had emigrated from Europe to escape lack of opportunities in his native Hanover, Germany, and avoid being drafted to fight in the Franco-Prussian War (1870–1871).

Aged 19, and with only a few dollars in his pocket, he disembarked in New York. Like many other immigrants with high hopes of his new homeland, Berliner was ambitious and hardworking but with little in the way of qualifications. His formal education in Hanover had stopped at age 14, and his subsequent work experience was in the family fabric store. Berliner first settled in Washington, DC, where he worked in a dry-goods store, but his interests in what would one day become "electronics" led him to attend evening classes in physics and electrical engineering at the Cooper Institute in New York, starting in 1873. In 1877, when he was working for the Bell Telephone Company, he filed his first patent for a carbon telephone transmitter, which was in essence the very first "microphone."

Thomas Edison, loser of the first "format war."

Although in some ways technically superior, the cylinder made way for the disk.

Into the groove

In technological format wars, as we shall see when we come to the entry on the VCR (page 184), it is not a foregone conclusion that the best technology wins: Price, ease of manufacture, product design, marketing and many other factors may, in the end, be far more important for the consumer than technical excellence. In terms of sound recording and reproduction, the disc had no inherent superiority over the cylinder. On the contrary, technically a cylinder had the advantage of a constant linear velocity, while with a disc the velocity slowed as the stylus approached the center. In addition, Edison's phonograph was much better than Berliner's first gramophones.

"[Berliner] began experimenting with discs on which sound could be engraved laterally instead of vertically on cylinders and this soon led to a photoengraved record that could be played back through a stylus and diaphragm reproducer."

Billboard, September 15, 1973

BERLINER "GRAMOPHONE"

SOUND RECORDING

- 1857   Phonoautograph
- 1877   Paleophone
- 1877   Phonograph
- 1881   Graphophone
- 1887   Gramophone

Not only was the sound reproduction superior, but the phonograph could also both play back and record, with a special device to shave off the wax surface of the cylinder so that it could be re-used. The disc, however, was much easier and cheaper to mass-produce by stamping, and it also had the advantage that it took much less space to store and ship. Berliner's first discs were 5 and 7 inches (12.7 and 17.5 cm) in diameter with music on only one side, but these were replaced at the beginning of the twentieth century by what became the standard double-sided 10-inch (25.4-cm), 78-rpm disc.

Berliner demonstrating his gramophone disc recording equipment.

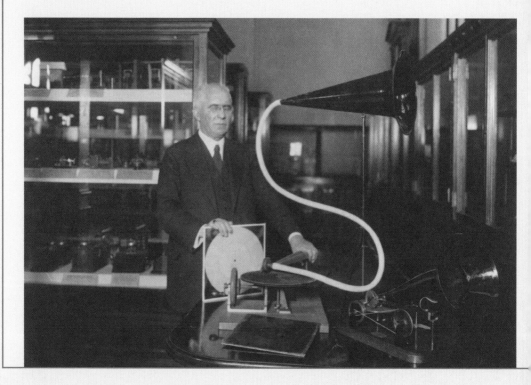

THE BERLINER "GRAMOPHONE"

THE GRAMOPHONE DISC

The key feature of the gramophone is not a component of the machine itself but the disc that it was designed to play. The disc was not technically superior to the cylinder but it was cheaper and easier to make and took much less space for storage. From 1895 until the 1940s, when vinyl was introduced, discs were made of brittle shellac mixed with pulverized rock and dyed black.

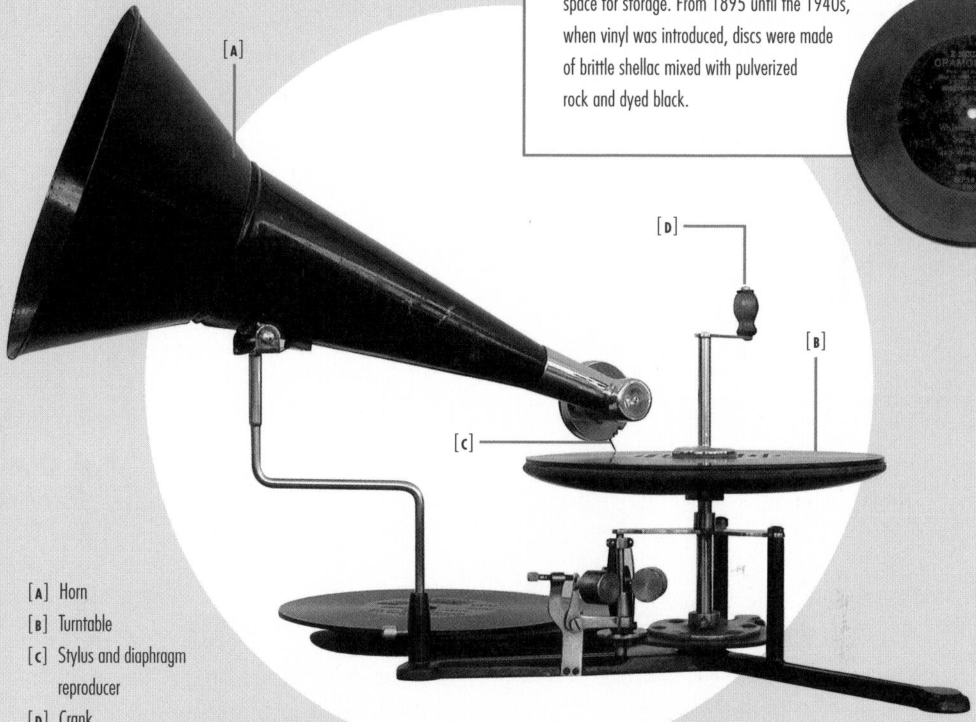

[A] Horn
[B] Turntable
[C] Stylus and diaphragm reproducer
[D] Crank

The earliest versions of the Berliner gramophone, made in 1889, were extremely simple devices that were sold in kit form and marketed as children's toys. The gramophone's most complex technical components were the discs it played and the diaphragm reproducer that converted the vibration picked up from the disc groove by the stylus into sound. The only form of sound amplification was provided by a metal horn that was attached directly to the stylus and reproducer. As can be imagined, the sound quality of the early models was extremely poor. The turntable was hand driven, but later gramophones were fitted with a spring-powered wind-up mechanism that did away with the need for an operator to turn the crank and also ensured that the disc would play at a constant speed.

17

Designer:
Léon Bouly

LUMIÈRE "CINÉMATOGRAPHE"

Manufacturer:

Lumière Company

Industry

Agriculture

Media

Transport

Science

Computing

Energy

Home

1895

The Lumière Cinématographe was an extraordinary machine: camera, processor and projector all in one. Although it was not the first device to capture movement on film, in the hands of the Lumière brothers, the Cinématographe established many of the conventions of modern cinema and by extension of all later visual media, including television and video.

Lights, camera, action!

Like many other key inventions of the Second Industrial Revolution, the motion picture camera and projector are traditionally credited to a single inventor—though in this case, a duo: the Lumière brothers, Auguste (1862–1954) and Louis (1864–1948). But like the gramophone, rotary dial telephone and light bulb, they owe their development to the work of many individuals going back centuries. The idea of projecting an image onto a screen with a camera obscura dates back to antiquity. However, for most film historians, the history of modern filmmaking begins with simple animation devices such as the "phenakistoscope" and "zoetrope." It is when these devices were combined with photographs and a light source, as in photographer Eadweard Muybridge's (1830–1904) "zoopraxiscope," that we begin to approach the experience of film projection.

The Cinématographe could be used both as a camera and a projector.

The history of early film technology features a name that readers will be familiar with from earlier entries (and their schooldays): Thomas Edison (1847–1931). Edison had a hand in many of the major inventions of the last quarter of the nineteenth century, although as with the gramophone and electricity generation, with the "Kinetoscope," he did not get it quite right.

Although the device introduced the 35-mm film format that would become the industry standard, it was not designed to project onto a screen but to be viewed by one person, like the later penny-in-the-slot machines that showed risqué movies in funfairs and seaside arcades. It was a Frenchman, Léon Bouly (1872–1932), who first designed the "Cynématographe Léon Bouly," which would not only capture moving images but also project them onto a screen. Unfortunately, Bouly did not have the funds to market his invention or even keep up the payments on his patent, which was acquired by the enterprising Lumière brothers, the owners of a large photography business in Lyon, France.

Lumière (but not son)

The Lumière brothers improved Bouly's Cynématographe and patented their own design, the "Cinématographe" in 1895. There is some dispute as to whether they saw Edison's Kinetoscope in Paris in 1894 before finishing their own machine, but for once, there was no patent infringement litigation; hence, the dispute is probably more to do with national pride as to who—the Americans or the French—invented cinema. However, watching Edison's Kinetoscope was an individual experience more akin to watching television or surfing the net, and it was the Lumière brothers who can claim the first movie-theater experience.

MOVING PICTURES

1832	Phenakistoscope
1834	Zoetrope
1879	Zoopraxiscope
1889	Chronophotographe
1891	Kinetoscope
1894	Mutoscope
1892	Cynématographe
1895	Cinématographe

Although they pioneered cinema, the Lumières abandoned the medium to concentrate on still photography.

An advertising poster for the first "cinema" screenings.

CINÉMATOGRAPHE LUMIÈRE

After several private screenings in 1895, the brothers held their first public performance in the basement of the Grand Café, Paris, on December 28, charging an admission fee of one Franc (around $4 at present prices). The program consisted of ten movies each running between 38 and 49 seconds, including the documentary subjects "Workers leaving the Lumière Factory," "Fishing for Goldfish," and "Blacksmiths," as well as an attempt at a comic short: "The Gardener, or the Sprinkler Sprinkled." In 1896, they took the Cinématographe on a world tour, visiting India, the U.S., Canada and Argentina. Despite their commercial success both in selling Cinématographes and projecting their short movies, the brothers thought that the cinema was "an invention without any future." They turned their backs on the film business to concentrate on their first love, still photography. In 1903, they patented their own color film process: "Autochrome Lumière."

"The composition and function of this lightweight 16-pound hand-cranked camera performed a threefold task: filming, printing, and projecting motion pictures [....] Thus the operator could shoot footage in the morning, process the film print in the afternoon, and then project it to an audience that same evening." R. LANZONI, FRENCH CINEMA (2004)

THE LUMIÈRE CINÉMATOGRAPHE

[A] Crank
[B] Lens
[C] Viewfinder

The Cinématographe is an extraordinary three-in-one device that could be used to film, process and project movies (something not repeated until the "Pico" projection system of 2003). In filming mode, the operator turned the crank at two revolutions per second to feed the perforated film stock (with two and not four holes as in standard 35-mm film) past the shutter at a rate of 16–18 frames/sec.

"The Cinématographe was notably portable and convertible as a camera, projector and film printer, so that the Lumière operators had a striking success in conducting pioneering exhibitions as they dispersed around the globe and sent the results of their filming back to Lyon."

DEAC ROSSELL, *LIVING PICTURES: THE ORIGINS OF THE MOVIES* (1998)

Cutaway of the Cinématographe showing the internal mechanism.

Part of the hand-cranked eccentric cam mechanism that drew the film through the Cinématographe.

The Cinématographe, like the still plate cameras of the day, could only be used outdoors in full daylight because the artificial lighting of that time was not strong enough to permit indoor or nighttime filming. It was also very limited in terms of the length of movies that could be filmed. The Lumière Brothers' features averaged at less than one minute each.

KEY FEATURE:
THE ECCENTRIC CAM MECHANISM

As the operator turned the handle, the Lumière brothers' patented eccentric cam mechanism converted the rotation into vertical motion, drawing the film past the shutter. The cam was attached to a flexible frame that had two pins that passed through the two perforations in the film stock, threading it through at the correct speed of 16 frames/sec. As the projector was hand-cranked, it demanded a certain amount of skill on the part of the operator to project the film at the right speed.

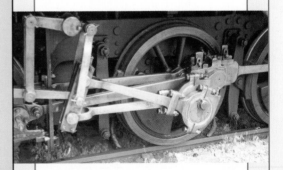

The eccentric cam mechanism was also incorporated into the design of trains.

This remained the standard frame rate until the advent of talking pictures when it was increased to 24 frames/sec. In order to make positive projection copies from a developed negative, the operator fed through an unexposed film together with the negative with the camera pointing at a uniform light source. In projection mode, the movie was run through the Cinématographe and projected onto the screen with the aid of an external light source.

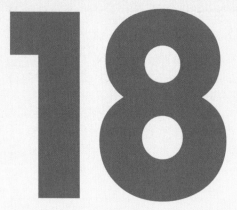

18

Designer:
Guglielmo Marconi

MARCONI RADIO

Manufacturer:
Wireless Telegraph & Signal Company

Industry
Agriculture
Media
Transport
Science
Computing
Energy
Home

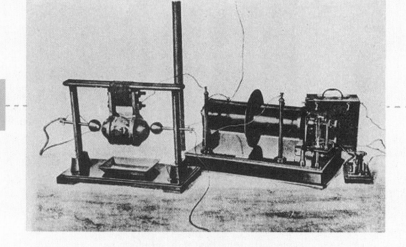

1897

Like the invention of the telephone and the light bulb, the history of radio is a minefield of claims and counter-claims. Most historians now agree that Marconi, though he commercialized radio, then known as "wireless telegraphy," was not its inventor. He built on previous theories and experiments and assembled and improved existing components to create a viable wireless communication system. Hence, though he should not be remembered as the man who invented radio, he was the brilliant entrepreneur who sold it to the world.

How radio helped sink the *Titanic*

On April 10, 1912 (for the benefit of readers who somehow managed to avoid seeing James Cameron's 1997 blockbuster), the RMS *Titanic* set sail on her maiden voyage from Southampton, England, bound for New York. Although she was short on lifeboats, she was equipped with all the latest technology, including two Marconi radios that could be counted upon to relay early warnings about the ever-present hazard to shipping in the springtime North Atlantic: icebergs. Acting on information received, the *Titanic*'s captain changed for a more southerly course. Unfortunately, he was steaming directly toward yet more icebergs. During April 14, *Titanic*'s radio operators received several iceberg warnings but because they were Marconi employees, and the company was paid to provide a lucrative wireless telegraph service for the first-class passengers, they did not relay the warnings to the bridge.

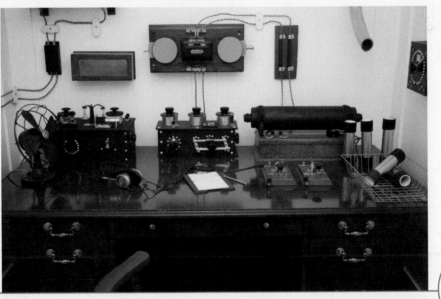

An original 1912 Marconi radio set identical to the ones onboard the RMS *Titanic*.

At around 11.40 pm, *Titanic* struck an iceberg, fatally holing her hull. Although radio had failed to avoid the disaster, it could have ensured the timely rescue of the passengers and crew. Although a number of ships received *Titanic's* "CQD" and the new "SOS" distress calls, none of those responding were close enough to reach her before she sank. The SS *Californian* was the closest but had stopped for the night because of the ice. She saw the *Titanic's* distress flairs, but her radio operator was asleep and her transmitter turned off, so she failed to get the distress call or respond. One story goes that *Titanic's* radio operator had upset the *Californian's* operator by refusing to take an earlier iceberg warning because he was too busy with passenger traffic. Tragically, instead of saving the *Titanic*, the radios on board contributed to her sinking with the loss of 1,517 lives.

RADIO

—1886   Hertzian waves

—1890   Branly coherer

—1893   Tesla demonstration

—1894   Lodge demonstration

—1894   Bose demonstration

—1894   Popov coherer

—1897   Marconi radio

"Marconi displayed the traits of an entrepreneur, had a knack for getting publicity, and seemed to proceed well in the commercialization of new wireless technology."

J. KLOOSTER, *ICONS OF INVENTION* (2009)

Going wireless

Until the 1870s when the SS *Great Eastern* successfully laid thousands of miles of undersea cables, the only means of communication between far-flung regions of the world was by steamship. Sending a message between London and New York took a week or more, and mail to Australia could take months, depending on the route the mail boat took. Wired telegraphy transformed the world, linking the continents for the first time, but it had one major drawback: it could not be used from ship to shore or ship to ship. As soon as the wired telegraph had been perfected, the race was on to find a way of communicating wirelessly.

The theory of communicating wirelessly via electromagnetic waves had been formulated by James Clerk Maxwell (1831–1879) in 1873 and confirmed by practical experiments by Heinrich Hertz (1857–1894) who could probably have claimed to be the first person intentionally to transmit and receive radio waves, later named "Hertzian waves" in his honor. However, he himself saw no use for his discovery. Shortly after his successful demonstration, he said: "I do not think that the wireless waves I have discovered will have any practical application."

Others, including Nikola Tesla (1856–1943) and Guglielmo Marconi (1874–1937), disagreed and saw in his work the key to a commercial system of wireless transmission for use on ships. Tesla, who was later recognized by the U.S. as the inventor of radio, gave demonstrations of his radio transmitter and receiver in Philadelphia and Chicago in 1893; several others in the U.S., UK, India, and Russia also managed to transmit radio signals before Marconi, but none of these gifted inventors managed to commercialize their discoveries. Another key advance in the history of radio was the development of a detector for radio waves, known as a "coherer" by Édouard Branly (1844–1940) in 1890.

Guglielmo Marconi was the brilliant entrepreneur who sold the radio to the world.

Keeping the queen informed

The "wireless telegraph" that Marconi patented in 1897 was very different from the voice and music broadcasting that we know as radio today. The first voice transmissions were still a decade away, and commercial radio broadcasts began in the 1920s. Early radio communicated by Morse code—combinations of short (dots) and long (dashes) pulses representing the letters of the alphabet (e.g., S.O.S. = • • • - - - • • •). In 1894, Marconi began experimenting with equipment modeled on Hertz's laboratory apparatus. His first transmissions were made in the family home's backyard in Bologna, Italy, but he soon increased the range of his signals to over 1 mile (1.6 km). He initially offered his invention to the Italian government, but when they declined, he arranged a demonstration for the British General Post Office. In 1896 he succeeded in detecting a transmission 8 miles (12.9 km) away, proving the viability of wireless communication. He opened the world's first radio factory in 1898 in Chelmsford, England. A master of self-promotion, Marconi scored a major publicity coup in December 1898, when he set up wireless communication between the Royal Yacht, where the Prince of Wales was convalescing after an injury, and Osborne House, Queen Victoria's (1819–1901) residence on the Isle of Wight.

Guglielmo Marconi

1874–1937

THE MARCONI RADIO SPARK-GAP TRANSMITTER

[A] Aerial
[B] Spark gaps
[C] Induction coil
[D] Battery
[E] Telegraph key
[F] Coherer

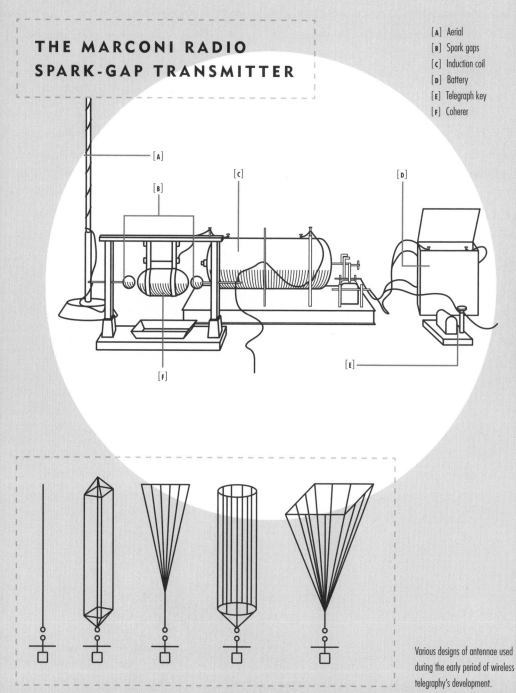

Various designs of antennae used during the early period of wireless telegraphy's development.

The earliest Marconi radios were spark-gap transmitters that used Morse code. The transmitter consisted of an induction coil, a Morse key, and a power supply. The first transmitters were battery powered, but Marconi later replaced this with an AC source and step-up transformer. The Morse key was connected to the alternator and transformer. When the operator depressed the key, current flowed into the induction coil that, in Marconi's early designs, acted as both the solenoid and step-up transformer. Twin contacts interrupted the flow of current to provide a series of pulses for the transformer. When the coil was magnetized, it attracted a metal bar attached to one of the contacts. This switched off the current, returning the contact to its original position. A parallel tuned circuit and a pair of spark gaps were connected to the transformer. When the Morse key was pressed it created sparks, so the voltage across the tuned circuit consisted of a series of damped oscillations at the frequency of the tuned circuit. The aerial and earth allowed the signal to be sent over significant distances.

This early spark transmitter, manufactured by the company Radiguet & Massiot in around 1900, was used on ships and had a range of about 6 miles (10 km).

KEY FEATURE:
COHERER

Early radio receivers used coherers to detect radio waves. Designed by Édouard Branly, the coherer consisted of a glass tube containing metal filings. When a high frequency current passed through the filings they stuck together, or "cohered," which reduced their electrical resistance. The coherer had to be tapped to separate the filings.

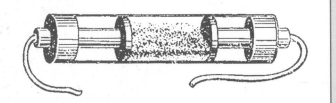

Édouard Branly's coherer was a vital part of early radio sets.

19

Rudolf Diesel

DIESEL ENGINE

Manufacturer:

Maschinenfabrik Augsburg

Industry

Agriculture

Media

Transport ■

Science

Computing

Energy

Home

1897

The steam engine drove the First Industrial Revolution, but it was extremely inefficient, making it wasteful of resources and expensive to run, while it also produced huge amounts of atmospheric pollution. Rudolf Diesel dreamed of creating an engine that would be truly efficient. His diesel engine, designed in 1892–1893 and successfully built in 1897, was one of the key inventions that contributed to the success of the Second Industrial Revolution.

Burning issue

The First Industrial Revolution was powered first by waterpower and then by steam. However, the stationary steam engine that drove the mills, mines, and factories, perfected by James Watt (1736–1819) and improved by George Corliss (1817–1888) was grossly inefficient, with a thermal efficiency of between 10 and 15 percent. This entailed a huge waste of resources and money, and also created atmospheric pollution on an epic scale. Although the environment was not the "burning issue" (pun intended) it is today, engineers strove to create an engine that would approach the efficiency of the ideal engine proposed by Nicolas Carnot (1796–1832) in 1824.

Carnot studied the engines of his day, including both steam and internal combustion (IC) engines, and he described an ideal "heat engine" that would work on the four-step cycle: Heat is transferred from a high-temperature source to expand the working substance (gas or liquid) and work the piston; as the piston and cylinder are perfectly insulated, they cannot gain or lose heat. The working substance expands, until the expansion causes it to cool; the cooling working substance transfers heat to the low-temperature "sink"; and compression of the working substance in the insulated engine causes its temperature to rise, returning it to the same state as in step one.

"When I began constructing my engine in the early nineties, the existing method was a total failure. The enormous pressures generated by the machine, the friction between the moving parts, the magnitude of which had never been seen before, forced me to minutely examine the stress on each single organ and to delve extensively into material science."

Rudolf Diesel quoted in Biodiesel (2008) by G. Pahl

DIESEL ENGINE

The first commercial four-stroke IC engine was the Italian Barsanti-Matteucci engine, patented in the UK in 1857. But IC only came into its own with the improvements made by the German engineers Nikolaus Otto (1832–1891) and Eugen Langen (1833–1895) in 1877. Four-stroke IC engines that used electrical ignition from spark plugs are still known as "Otto engines." Although much more efficient than steam engines, the Otto engine did not even begin to approach the thermal efficiency of the Carnot heat engine. A modern four-stroke engine only averages a thermal efficiency of around 30 percent.

The engineer vanishes

On the evening of September 29, 1913, the German engineer and inventor of the diesel engine, Rudolf Diesel (1858–1913) boarded a ferry from Antwerp bound for England. Although the First World War would break out in less than a year, there was nothing secret about Diesel's trip to the country that would soon be at war with his homeland. He was traveling to London for a routine meeting with the British manufacturers of his engines. He retired to his cabin at 10 pm, asking the steward to wake him at 6.15 am. The next morning, his cabin was found to be empty, and a subsequent search found no trace of him aboard. Ten days later, a body was found floating in the Channel by a Dutch fishing boat. The body was in such a poor state that the crew did not bring it aboard. Instead, they retrieved personal items from the body that might help in its identification. In October, Rudolf's family confirmed that the items belonged to the missing man.

The Barsanti-Matteuci engine, the first commercial internal combustion engine, would not have fitted into an automobile. This large vertical engine was designed to power heavy industrial plant and ocean-going ships.

Diesel's main biographer believed that, depressed and exhausted from overwork, Diesel had suffered a mental breakdown and taken his own life. But at the time conspiracy theories abounded in the British press. With war about to break out, it was suggested that German military intelligence had a hand in his death, in order to prevent him from passing on any more of his inventions to the British. Another more recent theory was that he was murdered on the orders of the oil industry, as he planned engines that would run on "bio-diesel," thus ending the lucrative monopoly of the petroleum companies on the production of fuel for IC engines. Despite the passage of almost a century, no evidence of a plot has ever been uncovered; hence, it is unlikely that the inventor's death was a mystery worthy of an Agatha Christie whodunit but merely suicide.

INTERNAL COMBUSTION ENGINE

- 1807   Hydrogen engine
- 1807   Pyréolophore
- 1857   Barsanti-Matteuci engine
- 1870   Marcus "car"
- 1877   Otto four-stroke engine
- 1879   Benz two-stroke engine
- 1882   Atkinson cycle engine
- 1897   Diesel engine

The diesel advantage

Diesel dreamed of creating an engine as efficient as Carnot's ideal engine, and although in theory diesel engines could reach 75 percent efficiency, in practice, they currently peak at 50 percent and average around 45 percent. This is still about 15 percent better than the efficiency of most Otto engines. The thermal efficiency of a diesel engine means, of course, that it uses less fuel for the same amount of work. Diesel oil itself is cheaper to produce from petroleum than gasoline, and can more easily be substituted with bio-fuels without any costly conversion of the engine. Although not a "green" fuel, diesel produces very little carbon monoxide (CO), which makes it ideal for use in mines and submarines. Because a diesel engine does not use a high-voltage ignition system, it has a much simpler overall design than a gasoline engine, which also makes it more reliable. The high pressures needed in a diesel engine mean that they have to be of much sturdier construction than gasoline engines, giving them a much longer working life. These advantages made diesel an obvious choice to replace steam and drive the heavy industrial and transportation plant of the Second Industrial Revolution.

THE DIESEL ENGINE

[A] Air inlet
[B] Fuel injection valve
[C] Cylinder
[D] Piston
[E] Connecting rod
[F] Crankshaft

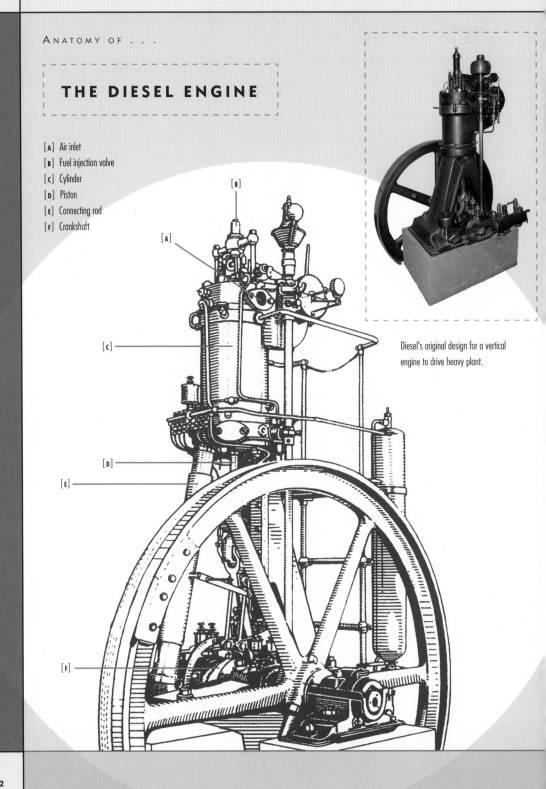

Diesel's original design for a vertical engine to drive heavy plant.

A diesel engine is an internal combustion engine that uses compression ignition instead of electrical ignition. Diesel's initial design was a vertical single-piston engine that drove a large flywheel. The engine cycle can be described in the four following strokes: (1) The intake stroke draws in air at atmospheric pressure into the cylinder through the intake valve; (2) with both valves closed, the compression stroke compresses the air to a pressure of 40-bar (4.0 MPa; 580 psi) and heats the air to 1,022°F (550°C); (3) at the top of the compression stroke, fuel is injected as a fine spray into the combustion chamber. The vapor is ignited by the heated compressed air in the chamber, causing the fuel droplets to burn at a constant pressure. The rapid expansion of combustion gases drives the piston downward, powering the crankshaft. In addition to allowing combustion to occur without a separate ignition system, the high compression ratio increases the engine's overall efficiency. A major problem in Otto engines where air and fuel are mixed before entry into the cylinder is the danger of premature ignition, which can wreck an engine. This is not a problem in a diesel engine, in which fuel does not enter the cylinder until shortly before top dead center (TDC); and finally (4) during the exhaust phase the waste burnt products are expelled via the exhaust.

KEY FEATURE:

AIR-BLAST INJECTION

Diesel's original engine used air-blast fuel injection to introduce atomized fuel into the cylinder with compressed air through a nozzle. A pin valve operated by the camshaft opened the nozzle and started fuel injection before TDC.

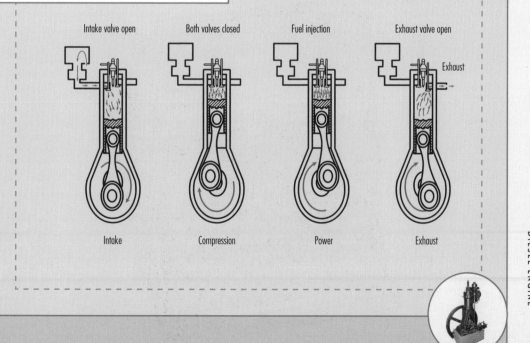

Intake valve open	Both valves closed	Fuel injection	Exhaust valve open
			Exhaust
Intake	Compression	Power	Exhaust

20

Franz X. Wagner

UNDERWOOD
NO. 1 TYPEWRITER

Manufacturer:

Wagner Typewriter Co.

Industry

Agriculture

Media ▪

Transport

Science

Computing

Energy

Home

1897

The invention of the typewriter marks the beginning of the "keyboard age." Although not the first "typewriter," the Underwood No. 1 incorporated most of the features that would become standard on manual typewriters until they were rendered obsolete by word processors and computers. Although now quaint antiques, in the late nineteenth century, they were the cutting-edge business technology that transformed office life just as much as the PC did a century later.

The "Writing Ball," 1870

The advent of the keyboard age

I want you to imagine an era during which everything had to be written painstakingly slowly in long hand—and not with a ballpoint or marker pen—but with a natural quill or metal-tipped pen dipped in an inkwell. Although I've never perfected my touch-typing, I can manage around 50 words per minute on my laptop, while a trained touch typist will work at speeds in excess of 120 wpm. But as the pace of industry, communication and transport accelerated during the Industrial Revolution, the business of commerce, finance and government was conducted at the stately handwritten pace of 20–30 wpm. By the mid-nineteenth century, the typewriter was an invention long overdue.

The history of the modern typewriter begins with the "Typographer" of 1829. This American device did not use a keyboard but a dial and was even slower than writing by hand. The 1855 Italian *Cembalo scrivano* ("Harpsichordwriter") was a strange hybrid of typewriter and piano,

and though much admired at the time was never commercialized. The first commercial typing device was the "Writing Ball," designed by the Danish pastor Rasmus Malling-Hansen (1835–1890) in 1870. In his first design a metal hemisphere studded with letter keys tracked over paper attached to a cylinder. Although ingenious, the Writing Ball was no match for the "Type-writer," designed by the American inventors Christopher Sholes (1819–1890) and Carlos Glidden (1834–1877) in 1868, and manufactured from 1873 by E. Remington and Sons, who would become one of the world's leading typewriter manufacturers. Although at first glance, the boxy Type-writer with its QWERTY keyboard resembles a modern manual typewriter, it lacked several important features: it had no shift key, so could only type in upper-case letters, and because the type bars struck vertically, the typist could not see what he was typing until the carriage return scrolled the paper up.

UNDERWOOD NO. 1 TYPEWRITER

I can see clearly now

Although it seems obvious to us now that there are distinct advantages for typists to see what they are writing as they type, the machines made at the end of the nineteenth century were "blind" writers, either because the mechanism was above the paper or the levers struck upward. Although not the first to design a visible typewriter design, Franz Wagner's (1837–1907) "rocker gear" of 1890 proved to be the most reliable mechanism for allowing the typist to see his work. Originally from Germany, Wagner emigrated to the U.S. in 1864. Trained as a mechanical engineer, he patented several inventions, including the first water meter, before turning his attention to improving the typewriter.

Although we have Wagner to thank for the design of the world's bestselling manual typewriters, as is often the case in the world of invention, he himself did not have the business acumen to capitalize on his discoveries. In 1895, he asked John T. Underwood (1857–1937), president of an office supplies firm that manufactured typewriter ribbons and carbon paper, for the backing to develop his invention. After Remington had begun to produce their own typewriter ribbons, Underwood riposted that he would manufacture his own typewriters to challenge the industry leader, the Remington Standard. He immediately recognized the strengths of Wagner's visible-type design, which went into production as the Underwood No. 1 in 1897. The first two generations of Underwoods bear the name "Wagner Typewriter Co." discreetly on the back of the machine in addition to the much larger "Underwood" on the front. However, any reference to Wagner was dropped in 1901, after he was forced to sell his patents outright to Underwood. By 1920, all rival typewriter designs had disappeared, and manufacturers worldwide imitated the Underwood's front-typing visible design.

TYPEWRITER

1829	Typographer
1854	Chirographer
1855	*Cembalo scrivano*
1870	Hansen Writing Ball
1873	Type-writer
1878	Remington Standard
1897	Underwood No. 1

The 1829 "Typographer" was slower than writing by hand.

The 14-ton typewriter

After a shaky start, the typewriter became the indispensable smart phone and computer tablet of its day—marketed with all the hype that marketing departments could devise. In 1915 Underwood created a 14-ton typewriter for the Panama-Pacific Exposition held in San Francisco. This metal monster stood 18 ft (5.4 m) high by 21 ft (6.4 m) wide and was a fully functioning replica of the Underwood No. 5, the world's bestselling manual typewriter, which could be operated by remote control. The type bars weighed 45 lb (20.4 kg) each, typing on a sheet of paper 9 x 12.5 ft (2.7 x 3.8 m).

But stunts aside, the typewriter, like the safety bicycle, played a significant role in the emancipation of women and their entry into the clerical workforce. Until the mid-1870s, the office was a predominantly male environment. Women either remained at home or worked in retail or factories. In addition to meeting the existing need for a faster means of writing, the typewriter created new roles in the office: the secretary, stenographer, and typist, which were predominantly performed by women, in part because they were ready to accept much lower wages than men. By 1900, three-quarters of clerical workers in the U.S. were women.

By 1920, the Underwood design had been adopted by typewriter manufacturers across the world.

The Underwood No. 5 was one of the most successful manual typewriters ever built.

"The Underwood No. 1 [...] was considered to be the first modern typewriter because, unlike earlier models, the type was fully visible as it was being typed."

A. Dewdney and P. Ride, *The New Media Handbook*, 2006

UNDERWOOD NO. 1 TYPEWRITER

THE UNDERWOOD
NO. 1 TYPEWRITER

[A] QWERTY keyboard
[B] Type bars
[C] Platen
[D] Carriage and
    carriage return
[E] Tabulator
[F] Space bar
[G] Shift key

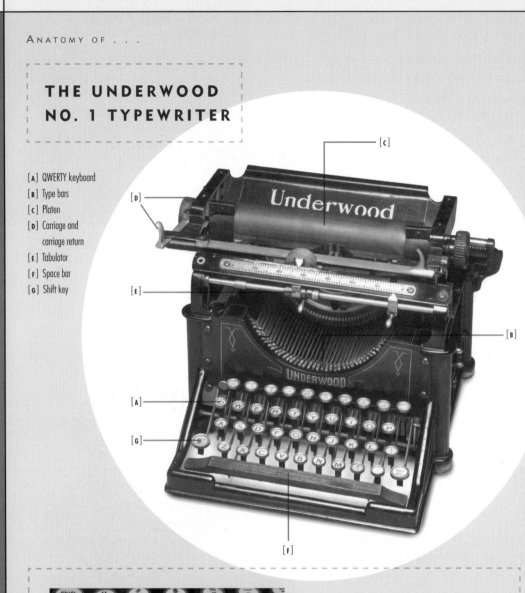

The Underwood came with a QWERTY keyboard and shift key.

KEY FEATURE:
VISIBLE TYPING AREA

The problem solved by Wagner was to devise a mechanism for the type bars that allowed the typing area to be visible while ensuring that the bars fell back into the correct place, and did not jam together while typing at high speeds.

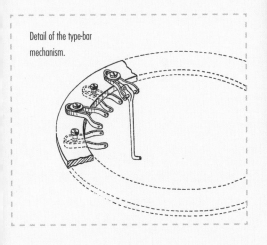

Detail of the type-bar mechanism.

A semicircular arrangement kept the type bars clear of the typing area and prevented jamming.

Pressing the levers raised the ink ribbon and moved the carriage forward.

The design of the Underwood No. 1 was built around the configuration of the type bars, or levers, that carried the letterforms. They were designed to hit the front of the platen (the roller holding the paper in place) rather than from underneath, as in earlier models. When not in use, the bars lay in a semicircular horizontal arrangement that did not obscure the typing area. Instead of having separate keys for upper- and lower-case letters, the operator selected the letters from a single QWERTY keyboard, and used the shift key to select for upper or lower case. The letter was impressed on the page by the letterform hitting an inked ribbon that was moved into place by the raising of the type bar. The operator could move the carriage without typing by pressing the space bar. When he reached the end of the page, he used the carriage return to move back to the beginning of the line and scroll the paper up. The Underwood was one of the first typewriters to have a built-in tabulator that allowed the operator to make neat columns. Although entirely mechanical, the Underwood had a lighter touch than other typewriters, which made it popular with clerical workers.

21

Designer:

Frank Brownell

KODAK "BROWNIE" CAMERA

Manufacturer:

Eastman Kodak Co.

Industry

Agriculture

Media ■

Transport

Science

Computing

Energy

Home

1900

Although initially marketed as a child's camera, the Kodak Brownie proved to be so useful that adults were soon buying it for their own use. The bestselling camera of all time, the Brownie's simple mechanism and operation revolutionized photography, allowing nonprofessionals to take family and holiday "snapshots."

"And here's one we took in..."

Today, we take it for granted that we can pick up our cell phones and take a picture of whatever is going on around us. But until the twentieth century, the business of taking photographs involved outsized plate cameras, with exposures of a minute or more, whose high cost put them out of reach of the average person. George Eastman (1854–1932), the founder of Eastman Kodak, was determined to transform photography from an elite profession to a pastime that anyone could enjoy. His first challenge was to do away with the complex plate technology that entailed the use of toxic chemicals. In 1884, he patented the first practical paper roll film. In 1885, he hired Frank Brownell (1859–1939) to help him design and build a camera for the new Kodak film.

Originally from Canada, Brownell had trained as a cabinet-maker. Starting with the holder for the film roll, designed in 1885, the two men went on to build the first Kodak camera, the "Kodak No. 1" in 1888. Although a breakthrough in the world of photography, the camera's sale price of $25 made it too expensive for most American pocketbooks. Eastman asked Brownell to come up with a camera that would not only be cheap but easy enough for a child to operate. The result was the Kodak "Brownie" No. 1 of 1900, which was initially aimed at children and retailed at $1.00. The camera was small, light, and easy to use, and the film roll, protected by an outer paper covering, could be loaded in daylight. It proved such a hit with both children and adults that in 1901, Eastman produced the Brownie No. 2, which cost $2.00, and remained in production until 1933.

STILL CAMERA

Camera obscura	4th century BCE
Niépce camera obscura	1816
Daguerreotype	1837
Wolcott camera	1840
Calotype	1841
Panoramic camera	1859
Kodak roll-film camera	1888
Kodak Brownie	1900

KODAK "BROWNIE" CAMERA

KODAK "BROWNIE" CAMERA

[A] Strap
[B] Lock
[C] Viewfinder lens
[D] Viewfinder
[E] Film advance
[F] Lens

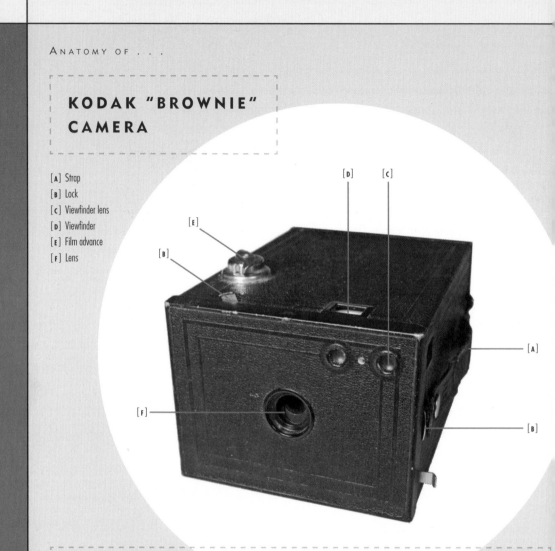

"Any school-boy or girl can make good pictures with one of the Eastman Kodak Co.'s Brownie Cameras, $1.00. Brownies load in daylight with film cartridges for 6 exposures, have fine meniscus lenses, the Eastman Rotary Shutters for snap shots or time exposures." KODAK ADVERTISEMENT, 1900

The early Brownies did not have any of the technical refinements of today's digital cameras. They were basically cardboard or wooden boxes (later replaced by aluminum) with a meniscus lens at the front and (on the Brownie No. 2) two viewfinders, one above the lens and one to the side above the shutter lever. Without a built-in flash or any means of adjusting the focus, aperture, or shutter speed, the Brownie could only be used outdoors in full daylight to photograph a static subject in the middle distance—ideal for that holiday snapshot of Aunt Ethel in Atlantic City. The Brownie took six frames of No. 120 Kodak film, measuring 2.25 in by 3.25 in (5.7 cm x 8.2 cm). After the operator had taken a picture, he had to wind the film forward manually by turning the knob attached to the spool. Because an outer layer of paper protected the film, the Brownie could be loaded in daylight, unlike earlier models that had to be loaded in a dark room to avoid exposing the film. The front of the box was hinged to allow opening to insert and remove the film roll. The Brownie was so easy to use that Kodak soon marketed it with the slogan: "You push the button, we do the rest."

The later Brownie Hawkeye model with an oversized flash, which was introduced in May 1949.

The Brownie replaced glass plates and toxic chemicals with easy-to-load paper roll film that could be sent to Kodak to be developed.

USE KODAK FILM V 120 OR 120

Kodak
VERICHROME FILM

USE KODAK FILM IN THE YELLOW BOX
It gets the Picture

PATENTED IN U. S. A.
1,169,882
1,176,329
1,494,719
1,548,116
1,613,365
1,620,304
1,771,483
PRINTED IN U.S.A. NO. 40,961

The Brownie's paper film roll was manually loaded onto the spool and moved forward by turning an external knob.

KEY FEATURE:
LOW COST

The outstanding feature of the Brownie range was value for money: the No. 1 retailed for $1.00 and the No. 2 for $2.00. Eastman also ensured that the cost of buying and processing the film for the camera was low: In 1900, a six-exposure film cost 15¢, paper negative film was 10¢, and processing was 40¢.

KODAK "BROWNIE" CAMERA

22

Designers:

Aleksandar Just
Franjo Hanaman

TUNGSRAM
LIGHT BULB

Manufacturer:

Tungsram

1904

The incandescent light bulb not only made the home and workplace a lot brighter, it also made them a lot safer as electricity was not combustible like kerosene and not explosive like lighting gas. Commercialized by Edison in 1880, the light bulb was perfected by the parallel work of Hungarian and American scientists working at the turn of the twentieth century.

The 67-year-old light bulb

I am indebted to EverythingWestport.com—the website dedicated to happenings in the town of Westport, MA—for an item of news that will gladden the hearts of those rearguard campaigners trying to save the incandescent light bulb from abolition and replacement by energy saving bulbs. In 2008, Ms. Elizabeth Acheson of Acheson farm, Westport, donated to the local historical society a light bulb that had been installed over the farm's porch in 1922, when the town was electrified, and remained in use until 1989. The sexagenarian bulb was a "Mazda," manufactured by General Electric (GE), with a six-loop tungsten filament.

The light bulb is another invention wrongly credited to Thomas Edison (1847–1931). Edison did not invent the incandescent bulb but successfully commercialized his design in 1880. The British scientist Humphry Davy (1778–1829) first demonstrated the principle of the light bulb in 1802, but it took almost 80 years for technology and materials science to advance sufficiently to make his invention commercially viable. The Edison bulb used a carbonized filament and glowed for an average of 40 hours—66 years and 363 days short of Ms. Acheson's Mazda. The major problem with early light bulb designs was the durability of the filament. Inventors experimented with a number of different substances, such as carbon and various metals, including platinum, but the one that was discovered to work best was tungsten.

"The modern form of incandescent lamp was a product of the twentieth century. The key development was the introduction of the tungsten filaments."

D. COLE ET AL, ENCYCLOPEDIA OF MODERN EVERYDAY INVENTIONS (2003)

TUNGSRAM LIGHT BULB

The tungsten Mazda first produced in 1909 by the Shelby Electric Co. and then by GE, however, was not the world's first tungsten light bulb, which we owe to the work of the Hungarian Aleksandar Just (1872–1937) and the Croatian Franjo Hanaman (1878–1941). The "Tungsram" light bulb (from a contraction of the names of the two metals used in the filament: tungsten and wolfram) went on sale in Europe in 1904. Although brighter and longer lasting than earlier filaments, it was perfected in 1909 by William Coolidge (1873–1975), Director of Research at GE, with his invention of "ductile tungsten."

LIGHT BULB

—**1802** Platinum filament

—**1809** Carbon arc lamp

—**1874** Carbon fiber filament

—**1878** Gas-filled globe

—**1880** Edison bulb

—**1904** Tungsram light bulb

[Right] A modern tungsten incandescent bulb
[Below] Early light bulbs with carbon filaments suffered from blackening and low luminosity.

ANATOMY OF . . .

THE TUNGSRAM LIGHT BULB

[A] Glass bulb
[B] Inert gas/vacuum
[C] Tungsten filament
[D] Contact wires
[E] Cap
[F] Electrical contact

Tungsten offered greater
durability and brightness than
any other material.

The incandescent tungsten light bulb is a marvel of applied engineering that was 100 years in the making. The perfectly blown glass bulb or globe originally protected a vacuum that prevented the filament from burning up in seconds rather than hours. However, there remained two major problems with light bulbs: blackening of the inside of the glass from the deposition of soot from the filament and low brightness. Although Just and Hanaman experimented with filling their Tungsram bulbs with inert gas to improve luminosity and reduce blackening, it was Irving Langmuir (1881–1957), a research scientist working for GE, who, in 1913, successfully produced a Mazda light bulb filled with the inert gas, argon. With a few other minor improvements, tungsten filament light bulbs have lasted until the twenty-first century. Although a vast improvement over previous methods of lighting: candles, oil and kerosene lamps, and gas lighting, incandescent light bulbs produce much more heat than light, and have a very low luminous efficiency: ranging from 1.9 percent for a 40-watt tungsten bulb to 2.6 percent for a 100-watt tungsten bulb—hence their imminent phase-out in most of the developed world.

The world's first tungsten light bulb was the
Hungarian-designed Tungsram.

23

Designer:
Almon Strowger

AUTOMATIC ELECTRIC CANDLESTICK TELEPHONE

Industry

Agriculture

Media ■

Transport

Science

Computing

Energy

Home

Manufacturer:

Automatic Electric Company

1905

It would be difficult to exaggerate the social and economic impact of the telephone after its commercialization in the last quarter of the nineteenth century. However, fixed-line telephony didn't achieve its full potential until the invention of the Strowger stepping switch made automatic dialing possible. With the Strowger Automatic Electric exchanges came the rotary-dial telephones that were standard until the touch-tone phones of the late twentieth century.

Almon Strowger

The much-contested invention

One of the features of the late nineteenth century was the incidence of acrimonious patent litigation, especially around key media and communication inventions. These disputes, which sometimes continued for decades, demonstrated how difficult it was to decide on the originality of an invention that was manufactured from existing components and based on established theoretical principles that often led to the development of similar devices simultaneously; it also highlighted the huge financial gains to be made from being the first to commercialize certain innovations. The invention of the fixed-line telephone is probably the most contested patent of the period, and arguments about who invented the phone can still trigger acrimony between the supporters of different claimants.

As soon as the first commercial wired telegraph entered service in the UK in 1839, transmitting Morse code, inventors began to work on how to transmit the human voice over its wire connections. Among those who can claim a hand in the invention of fixed-line telephony, we can include the Franco–Belgian Charles Bourseul (1829–1912), inventor of the "make-and-break" telephone in 1854, the German Johann Reiss (1834–1874), whose "Telephon" of 1860 was the first device to be called a "telephone," and the Italian Antonio Meucci (1808–1889), who filed a caveat for his 1854 invention of electromagnetic voice transmission at the U.S. Patent Office in 1871.

"You will never realize the true value of a perfect telephone service until you install the AUTOMATIC, UNMEASURED, UNLIMITED, AND SECRET SERVICE."

FROM A 1910 ADVERTISEMENT FOR A PHONE COMPANY USING AUTOMATIC ELECTRIC PHONES

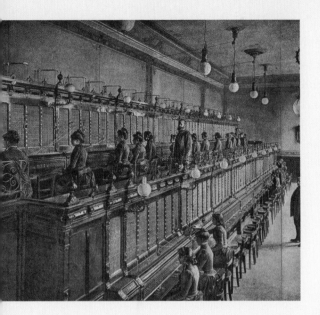

A pre-Strowger-switch exchange employed hundreds of operators.

TELEPHONE

The two main patent contenders for the telephone in the U.S. are, of course, Alexander Graham Bell (1847–1922) and Elisha Gray (1835–1901). Both Bell's patent and Gray's caveat were filed on the same day a few hours apart, and there were accusations of theft and patent fraud, as well as conspiracy theories to explain why Bell eventually won out over Gray. However, as seems usual in these disputes, the person who succeeded was also the first to have a workable device that could be taken to market.

"Girl-less, cuss-less, out-of-order-less, and wait-less."

At first telephone lines were installed on an individual basis, connecting two subscribers directly without any intervening exchange. To make a "call," one party whistled down the phone until the other party picked up the phone. This state of affairs, had it continued for any time, could have meant that our cities would have been covered in a dense net of telephone wires; however, the first telephone exchange modeled on the telegraph exchanges of the day entered service in 1878. You may have seen an early exchange in an old movie or a period drama. They operated very simply: The caller first contacted the operator, who replied: "Number, please."

The operator established the connection with pairs of jack cords that she plugged in manually into a switchboard panel. The system worked fine when there were few subscribers, but as soon as the number of telephone lines increased, and especially after the installation of the first long-distance lines, it was clear that a more rapid and efficient means of connecting phone users was required. Ultimately this would mean automation.

The story goes that in the late 1880s, Almon Strowger (1839–1902), a Kansas City funeral director, became convinced that a rival firm was stealing his business, because the owner's wife, who was the local telephone operator, was redirecting his calls to her husband. This spurred him to devise an automatic dialing system that would do away with the need for a human operator. He designed what would later become known as the "Strowger switch," the basis of the first automated telephone exchanges, and introduced the rotary dial telephone that remained in use for eight decades. When he opened his first exchange in LaPorte, IN, in 1892, serving 75 subscribers, he reportedly boasted that his Automatic Electric exchanges would be: "girl-less, cuss-less, out-of-order-less, and wait-less."

The Strowger switch was suitable for smaller exchanges, while larger city exchanges continued with human operators.

AUTOMATIC ELECTRIC CANDLESTICK TELEPHONE

Automated exchanges assured complete privacy to callers, and also increased the speed of telephony.

The immediacy and intimacy of the telephone call transformed personal relationships.

"My word! You do tickle me."

Private lines

The telegraph provided the first mass telecommunications system, but it had severe limitations: it was text only, messages were expensive and kept short, and they were not private, as the messages had to go through an operator who received the Morse code and transcribed them for delivery. Until the invention of the Strowger switch, the telephone also suffered from the latter drawback, as there was always the risk that the operator was listening to a private conversation.

The fear that his business was being sabotaged was what had driven Strowger to come up with the idea of the automated exchange, and the privacy of the system was undoubtedly a major selling point for business users who could speak to their customers with confidence and at much greater length than with the telegraph. Slowly, the telephone stimulated the development of new businesses. The introduction of automation also increased the speed of the telephone, which was another advantage to commercial users.

But perhaps the greatest impact was on social relationships. The telegraph was too expensive to use in anything else but an emergency, and the only other means for family and friends to stay in touch was by mail. The telephone provided an immediacy and intimacy that was the beginning of all later forms of social networking.

THE AUTOMATIC ELECTRIC CANDLESTICK PHONE

The original Strowger Automatic Electric telephones did not have a rotary dial. Anticipating the much later development of push-button technology, they had a button that the caller had to tap the correct number of times in order for the switches in the exchange to "step" and connect with the desired phone. In 1896, Automatic Electric introduced the first rotary dial phone, which remained in use until the late twentieth century. The candlestick design featured here dates from c. 1905. It featured the "knuckleduster" dial, with 10 holes for the digits: 0–9, and an eleventh hole for the operator for long-distance calls. Later models, such as those featuring the "Mercedes dial," had the standard ten finger holes. Modeled on a candlestick, the phone had the mouthpiece at the top of the stem, with the hook and detachable earpiece to the side. The wall-mounted ringer was a separate unit connected by wires to the base of the phone.

[A] 11-hole "knuckleduster" dial
[B] Mouthpiece (transmitter)
[C] Earpiece (receiver)

KEY FEATURE:
THE STROWGER SWITCH

The key feature of the Strowger system was not in the phone itself, but the design for an automated switching mechanism that would replace the operator in the telephone exchange. The switch had a ten-by-ten contact matrix to which the telephone lines were wired. When the caller dialed the number, an arm with an electrical contact would "step up" the rows until it reached the row corresponding to the first dialed digit. The process would be repeated for the subsequent digits until the connection was established. At the end of the call, the switch would be released and return to the start position.

A Strowger selector assembly, showing the contact matrix to which the telephone lines were attached.

24

Designer:
Childe Harold Wills

MODEL T FORD

Manufacturer:
Ford Motor Co

Industry
Agriculture
Media
Transport
Science
Computing
Energy
Home

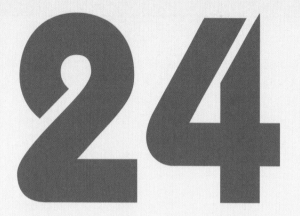

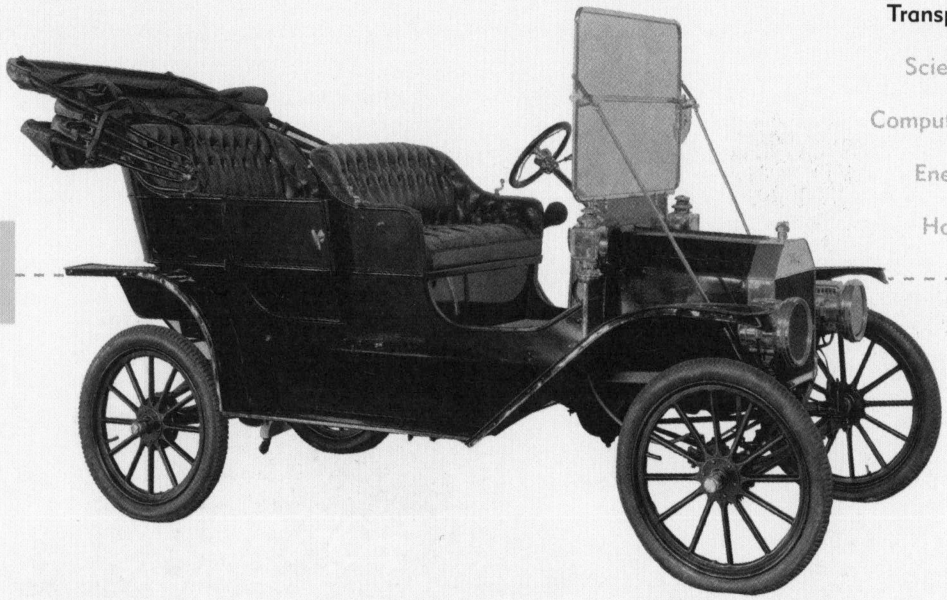

1908

Although not the first to manufacture automobiles in the U.S. or to use mass-production, Henry Ford created the world's first mass-market automobile. His success not only established Ford Motor Company as one of the industry leaders, but it transformed society in ways that Ford himself could never have anticipated. Although cars have given us economic development and freedom of movement, they have also brought pollution and gridlock.

The horseless carriage boom

In the closing decades of the nineteenth century, the biggest environmental headache for city planners in North America and Western Europe was not poor sanitation, slum overcrowding, or pollution, but the horse. By 1880, there were so many horse-drawn vehicles in the world's major metropolises that the city fathers of New York, Paris, and London were warned that by 1930 their streets would disappear under 9 ft (3 m) of horse manure.

In 1886, as if answering their prayers, Karl Benz (1844–1929) produced the Benz-Patent Motorwagen (Motorcar), the world's first commercial automobile equipped with a gasoline-powered internal combustion engine. With its high carriage seat and outsized rear wheels, the Motorwagen looked a lot like a buggy whose horse had gone astray mid-ride. But with this rather ungainly, clunky beginning, the Automobile Age had begun.

Henry Ford

"I will build a car for the great multitude. It will be large enough for the family, but small enough for the individual to run and care for. It […] will be so low in price that no man making a good salary will be unable to own one."

HENRY FORD, MY LIFE AND WORK (1922)

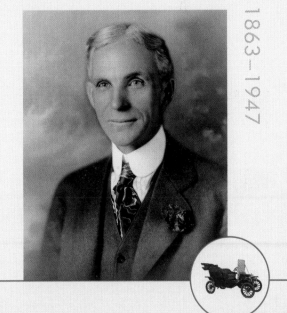

1863–1947

MODEL T FORD

The first decades of the auto industry were akin to the dotcom boom of the late 1990s. Hundreds of young idealistic inventors and engineers could see the tremendous potential of the automobile, but their dreams were outpacing the materials science and technology of the day and the road and gasoline infrastructure (gas was originally sold as a cleaning product by drugstores). They were also guilty of overestimating the size of the market. The first auto offerings were handmade in tiny quantities and sold to wealthy enthusiasts and hobbyists. Benz, the leading automaker of the period, only sold 25 Motorwagens between 1886 and 1893. Undeterred, hundreds of engineers and inventors went into business, experimenting with a wide range of engine technologies, including steam, electricity, and even early hybrids, alongside internal combustion, using an equally wide range of fuels. However, like any investment "bubble" growing from any new technology, after the initial excitement, many firms went bust, shattering the dreams of young entrepreneurs and losing investors their money.

Investing in the Ford dream

In 1903, the attorney Horace Rackam (1858–1933) went to his bank for advice about a potential investment. The investment in question was 50 ordinary shares, at $100 a share, in a new automobile company being set up by his friend and client, Henry Ford (1863–1947). The unnamed bank manager replied dismissively, "The horse is here to stay, but the automobile is only a novelty—a fad." Rackam, to his credit and considerable profit, ignored the advice and raised the $5,000 stake by taking out a loan and selling real estate holdings. In investing in the Ford Motor Company, however, he was taking a significant risk.

Henry Ford was the son of an Irish father, and a second-generation Belgian mother, who were farmers in Greenfield, near Detroit, MI. Henry's parents hoped he would take over the family farm, but he had other interests and ambitions. Aged 16, Ford went to Detroit to become an apprentice machinist. After another stint on the family farm, he got a job with someone with whom my readers will now be familiar: Thomas Edison (1847–1931), as an engineer in the Edison Illuminating Company.

Henry Ford's first auto attempt was
the Ford Quadricycle of 1896.

A Model T Ford coupe with "artillery" wheels.

Ford was promoted to Chief Engineer in 1893, but in his spare time, he experimented with the new internal combustion engines that were coming out of Europe. In 1896, he built his first vehicle, the Ford "Quadricycle"—more a motorized cycle than a true automobile.

In 1899, Ford left Edison and set up the Detroit Automobile Company, determined to crack the burgeoning auto market. He hired Childe Harold Wills (1878–1940), who would play a major role in designing later Ford models. The Detroit Automobile Company faltered, to be succeeded by the Henry Ford Company in 1901, and finally, in 1903, by the Ford Motor Company. It was then that Ford realized that rather than build high-performance racers and roadsters for the few, he should manufacture "a car for the great multitude."

U.S. AUTOMOBILES

Duryea Motor Wagon	1893
Ford Quadricycle	1896
Packard Model A	1899
Oldsmobile Curved Dash	1901
Cadillac Runabout	1902
Cadillac Model A	1903
Ford Model A	1903
Ford Model T	1908

MODEL T FORD

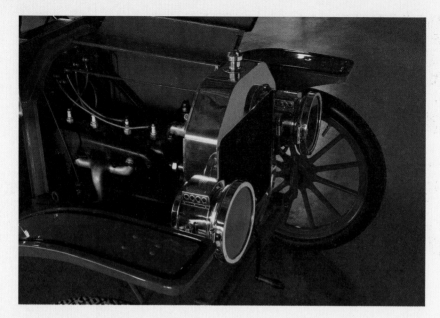

The front of the Model T with the hood up, showing the engine; the crank handle was used to start the engine.

The workingman's Ford

With a couple of business failures behind him, Ford risked everything—his reputation and all his own and his investors' money—on the first automobile to be built by the Ford Motor Company, the Ford Model A. Ford's gamble paid off, and he sold over 1,700 Model As, securing the future of the firm. His subsequent rise was nothing short of meteoric: In 1904, he founded Ford Canada; in 1906, Ford became the top-selling auto brand in the U.S. with close to 9,000 cars sold; in 1909, he founded Ford of Britain, opening the company's first overseas factory in Manchester, England, in 1911; and in 1913, Ford established itself in South America with Ford Motor Argentina.

Between 1903 and 1908, Ford and Wills designed nine models, each given a letter, though not all got past the prototype stage. In 1908 Ford introduced the Model T, which remained in production until 1927, with 15 million cars sold worldwide. Although Ford did not invent mass production (or the auto assembly line—Oldsmobile had done it before him in the U.S.), he introduced it on a much larger scale and applied it more rigorously in his factory at Highland Park. By 1914, the production time for the Model T had dropped from 12.5 hours to just 93 minutes.

THE MODEL T FORD

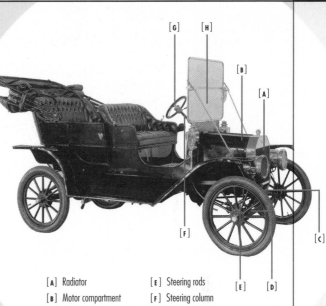

[A] Radiator
[B] Motor compartment
[C] Crank
[D] Front axle
[E] Steering rods
[F] Steering column
[G] Steering wheel
[H] Dash and windshield

The Model T looks like a basic automobile, with the usual arrangement of front-mounted engine, chassis, pedals, steering wheel, and four wheels, but looks can be deceptive. Because the Model T does not have a battery, the driver started the engine by hand-cranking a magneto, with a handle at the front of the radiator. Because the engine could "kick back," the driver had to cup the handle in the palm of his hand rather than grasp it to avoid a broken thumb. The choke was operated from a wire at the base of the radiator so that it could be used during cranking. Once started the four-cylinder inline engine had a top speed of 40–45 mph (64–72 km/h) with a fuel consumption of 13–21mpg (5–9 km/l). But the greatest difference from a modern car was in driving. The driver used three pedals (high and low gears,

KEY FEATURE:

PLANETARY GEAR TRANSMISSION

Although billed as a three-gear automobile, the Model T had two forward gears—low and high—and reverse. The main braking mechanism was an engine brake on the transmission. Like other parts of the car, the planetary gear transmission was made of advanced heat-tempered vanadium steel alloy.

reverse gear, and engine brake) and two levers (handbrake and throttle). To put the car into low gear, the driver put the handbrake into mid position or fully forward and pressed the left pedal down; to go into high gear, the driver pushed the lever forward and let the left pedal up.

Starting early engines with a crank often produced a "kick back."

25

James Murray Spangler

HOOVER SUCTION SWEEPER

Manufacturer:

Hoover Company

Industry

Agriculture

Media

Transport

Science

Computing

Energy

Home ■

1908

The Hoover Suction Sweeper, the first upright electric vacuum cleaner, was one of the many labor-saving devices that began to transform the life of women in the first decades of the twentieth century. A slow seller at first, the sweeper was marketed with one of the first home trial offers.

From pillowcases to riches

The Hoover story is a heart-warming tale, which, though more "small-town business to multinational corporation" than "rags to riches," can be held up as an exemplar of the "American dream." Unfortunately, the person who realized the dream was not the inventor of the world's first electric upright vacuum cleaner, James Murray Spangler (1848–1915), but his financial backer, business associate and cousin by marriage, William H. Hoover (1849–1932). That's why, in Britain, where the name Hoover is so closely associated with the vacuum cleaner that the two terms are interchangeable, we don't "spangler" the carpet but "hoover" it.

According to the Hoover website, Spangler worked as a janitor in Canton, OH. This is true, but it fails to mention that he was an inventor with several patents for agricultural machinery to his name. Unfortunately, he was not a good business-man, and he never managed to strike it rich from his inventions. Hence, in his late fifties and suffering from asthma, he was working as a janitor at the Zollinger Department Store. Because sweeping the store's carpet made his symptoms worse, he decided to make an electric sweeper that would suck the choking dust into a bag. Starting with a manual rotary sweeper, he added a sewing machine motor and leather belt to turn the sweeper brush and power a fan that blew the dust into a pillowcase. In 1907, having established the principle, he improved on the design, applied for a patent (awarded in 1908) and established the Electric Suction Sweeper Company.

"At a cost of less than one cent, you can thoroughly clean any room. Simply attach the wire to an electric light socket, turn on the current and run it over the carpet. A rapidly revolving brush loosens the dust, which is sucked back into the bag."

A "FREE HOME TRIAL" ADVERTISEMENT FROM GOOD HOUSEKEEPING, 1908

HOOVER SUCTION SWEEPER

From home trial to world domination

Unfortunately, Spangler's lack of business acumen let him down once more. He did not have the funds to mass-produce the sweeper. He demonstrated the machine to his cousin, Susan Hoover, who was so impressed that she told her husband about it. William Hoover owned a horse harness store in North Canton. But with the growing popularity of the automobile threatening his business, he was looking to diversify. He bought into Spangler's patent and invested in his company, which later took its new owner's name. Spangler continued to work for Hoover and to improve his invention, but he died suddenly in 1915, on the eve of his first ever holiday.

Although the suction sweeper was superior to its competitors, because, as the advertising slogan proudly boasted: "It beats, as it sweeps as it cleans!" early sales were slow. Hoover hit upon the idea of a free 10-day home trial, in which the customer only had to pay to return the sweeper if not completely satisfied. The marketing ploy worked and within the next decade, Hoover had established itself as a leading global brand. Hoover was fortunate that he was marketing a labor-saving device for the home just at a time when the role of women in society was being transformed. Labor shortages during the First World War had forced many more women to go to work, and social trends such as female emancipation and the appearance of alternative employment to domestic service made the vacuum cleaner extremely attractive to the hard-pressed homemaker of the 1920s.

VACUUM CLEANER

- 1868 Whirlwind
- 1876 Bissell carpet sweeper
- 1899 Pneumatic carpet renovator
- 1901 Puffing Billy
- 1908 Hoover Suction Sweeper

Before the suction sweeper, most homemakers used manual sweepers such as this 1881 Bissel.

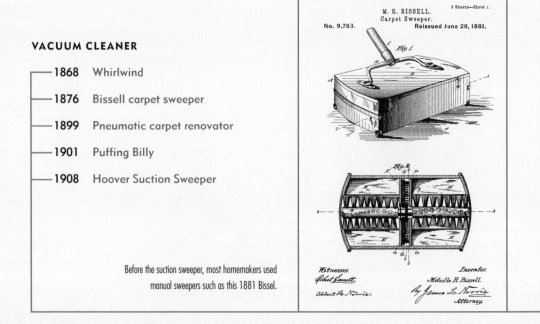

THE HOOVER SUCTION SWEEPER

[A] Bristle brush
[B] Fan
[C] Electric motor
[D] On/off switch
[E] Dust bag

Although promoted in period ads as "this little machine," the Hoover Model 0 was a 40-lb (18-kg) monster that must have given its operator a great workout. However, compared to the alternative—sweeping carpets by hand or hanging them outdoors to beat them—the suction sweeper must have looked like a marvel. And compared to earlier powered vacuum cleaners, which were mounted on horse-drawn wagons, they were amazingly compact. Spangler's original design for the upright Suction

Sweeper, although it got smaller, lighter, and more powerful remained largely unchanged right until the development of bag-free vacuum technology in the twentieth century. A brush at the front of the machine turned (and later beat), dislodging dirt from the carpet, and a fan powered by an electrical motor sucked it into a bag.

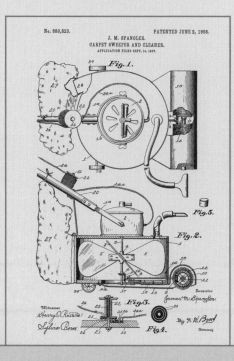

Illustrations from James Spangler's patent application, lodged in September 1907.

26

Designer:

Benjamin L. Holt

HOLT CATERPILLAR
COMBINE
HARVESTER

Industry

Agriculture ■

Media

Transport

Science

Computing

Energy

Home

Manufacturer:

Holt Caterpillar Company

1911

114

At the same time Henry Ford was revolutionizing U.S. industry with his car assembly line, Benjamin Holt achieved no less a far-reaching transformation of U.S. agriculture when he introduced the first self-propelled internal-combustion combine harvester. The rapid mechanization of agriculture led to the migration of agricultural workers to cities to seek new jobs.

Benjamin Holt

Cultures of consumption

For most of human history (and a good part of prehistory), the principal occupation of the vast majority of the working population was the production of food: principally grains of various kinds, such as wheat and barley in Europe and North America. It is estimated that in 1800, approximately 90 percent of the U.S. population worked on the land. Until the early part of the nineteenth century, cereal production depended entirely on the labor of humans and animals: Farmers plowed the land with the help of oxen, mules and horses, and they reaped their harvest by hand with scythes and threshed it using methods that had changed little in millennia. The First Industrial Revolution affected every area of society, and agriculture was no exception. Mechanization began with a reaping machine patented in England in 1799. However, the first major breakthrough in harvesting technology came in 1835 when Hiram Moore (1801–1875) designed the first horse-drawn combine harvester.

Two countries led the way in the development of agricultural machinery in the late nineteenth century: Britain, the period's technological superpower, and the United States, with a huge and growing agricultural sector. During the 1880s and 90s firms on both sides of the Atlantic developed new agricultural machinery, including the Holt brothers, whose company manufactured wooden carriage wheels near San Francisco. The youngest brother, Benjamin Holt (1849–1920), was acknowledged to be the brightest, most able, and the most mechanically gifted. In 1886, he designed a link-belt combine harvester that used flexible chain belts attached to the wheels to power the machine, and in 1891, a combine harvester with self-leveling technology that could harvest slopes. The drawback was that the new combines were so large that they needed up to 20 horses or mules to draw and power them. The solution was to find an alternative power source, which at the time meant steam.

> "The development of cereal-grain harvesting machinery, notably the reaper and threshing machine, was a highlight of the nineteenth century. Mechanization was seen as the prime source of advantage in wheat production."
>
> AMERICAN AGRICULTURE IN THE TWENTIETH CENTURY (2002) BY BRUCE GARDNER

The power of traction

Holt's first steam-powered "tractor" was a 24-ton (22,000-kg) monster mounted on huge iron wheels, built in 1892. Although slow, it could haul 50 tons (45,000 kg) and harvest a field more cheaply than a horse-drawn combine. However, the weight of the tractor meant that it was not suitable for softer ground. In 1903, Holt traveled to England to study the latest developments in agricultural technology.

Although a patent dispute would ensue between Holt and a British inventor, after Holt's return to the U.S., he came up with an invention that would shape the future of his company and define a whole new class of machines: the caterpillar track—still used today on heavy construction plant and military vehicles. The invention was so important to the firm that in 1911 Holt decided to rename it the Holt Caterpillar Company (better known as CAT today). That same year, Holt made another breakthrough when he designed the world's first combine harvester entirely powered by internal combustion (IC). The advantages of IC over steam were reduced weight and size and improved fuel efficiency. The innovations introduced by Holt would radically reduce the need for agricultural workers, who migrated to find jobs in the new industrial cities.

COMBINE HARVESTERS

- 1799   Reaping machine
- 1826   Bell reaping machine
- 1831   Virginia Reaper
- 1835   Moore horse-drawn combine harvester
- 1872   Reaper-binder
- 1911   Holt combine harvester

Holt Caterpillar Company made an important contribution to the Allied war effort during the First World War; their tractors were used to move heavy equipment and crew.

THE HOLT CATERPILLAR COMBINE HARVESTER

[A] Gasoline engine
[B] Thresher
[C] Crawling tread
[D] Straw walkers
[E] Tiller wheel

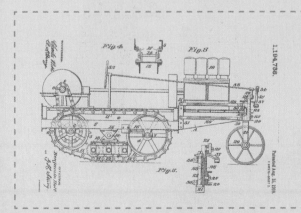

Although it was similar to a horse- or steam tractor-drawn design in its basic operation, the Holt Caterpillar combine harvester was the first to be powered entirely by an internal combustion engine. The crop entered the machine through the header, located to the side of the machine and not the front as in a modern design, and into the threshing mechanism where the grain was separated from the stalks. The grain passed through sieves and was stored. When the storage bin was full it had to be emptied into a trailer with the unloader. The chaff, made up of the stalks and other waste, was removed by straw walkers and dispersed over the ground by a spreader.

Benjamin L. Holt was highly protective of his designs and was not afraid to confront competitors over perceived copyright infringements.

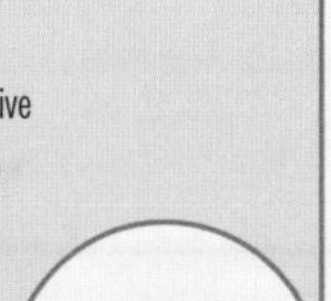

27

Designer:
Alonzo Decker

BLACK & DECKER
ELECTRIC DRILL

Manufacturer:

Black & Decker Manufacturing Company

Industry

Agriculture

Media

Transport

Science

Computing

Energy

Home

1917

Once AC supply had become commonplace, and after the invention of small practical electric motors, it was not only electric lighting that entered the home, but a succession of labor-saving devices powered by electricity. Among the first generation was the Black & Decker electric drill.

Drill at your own risk

It is not only drilling for oil in the Gulf of Mexico that is hazardous to the nation's health. According to health-and-safety statistics from all over the developed world, literally thousands of home improvers (DIY enthusiasts to the Brits) suffer serious and sometimes fatal injuries when using power tools. Top of the list comes the homeowner's trusty companion for almost a century, the power drill. Starting with those who fall off the ladder, and followed by those who drill into their own bodies, we come to those who, like this writer, drilled into a service duct that, in my case, carried the house's electrical supply. Home-improvement casualties notwithstanding, the introduction of the first handheld electric drill in the early twentieth century began a revolution in construction, maintenance and home improvement, and heralded the development of hundreds of domestic versions of industrial machine tools.

Hand drills have existed since prehistory, and water- and wind-powered versions are known from antiquity. The Industrial Revolution witnessed the invention of high-precision machine tools for industry, which included large, fixed steam-powered drills. But until the second decade of the twentieth century, there were no portable power tools designed for the small tradesman or domestic user. The invention of the first practical electrical motor in 1873, and the provision of a dependable electricity supply in the 1880s, created an entirely new market for domestic products powered by a household's main electricity supply, such as the vacuum cleaner. With the American tradition of homeowners building and maintaining their own homes, inexpensive, portable power tools represented an amazing commercial opportunity for any company that could get the right design to market. With its versatility and broad applications, the world's first portable power tool was the electric drill.

> "Electric drills and other portable hand tools became an important part of household kits in much of the industrialized world in the second half of the twentieth century. The basic invention was much earlier by men whose names are associated with the world's preeminent manufacturer of power tools for domestic markets, Black & Decker."
>
> AN ENCYCLOPEDIA OF MODERN EVERYDAY INVENTIONS (2003) BY D. COLE ET AL

The dream partnership

S. Duncan Black (d. 1951) and his partner Alonzo G. Decker (1884–1956) met in 1906 when they were both working at a firm manufacturing printing equipment for the telegraphic industry. Although they were both model employees, there was little hope of promotion in the firm, and in 1910, they decided to set up the Black & Decker (B&D) Manufacturing Company in Baltimore, MD. The two were extremely short on cash, and Black had to sell his most treasured possession, his car, to raise his share of the initial $1,200 stake, with a further $3,000 obtained from backers. If Decker was the technical genius in the partnership,

Black, who was a natural-born salesman, was the business brains of the outfit. Initially, B&D made products to order for other companies, but in 1917 they brought to market the first of a long line of B&D products, a portable air compressor. The device sold moderately well, but the company would have gone under had it not been for the release later that year of Decker's design for the first portable electric drill. With its original features, the B&D drill was an instant worldwide hit, selling in Europe, Australia and Japan. By the end of the decade, the company had several factories and annual sales in excess of $1 million.

By the early 1920s, the Black & Decker brand had achieved an international profile.

THE BLACK & DECKER ELECTRIC DRILL

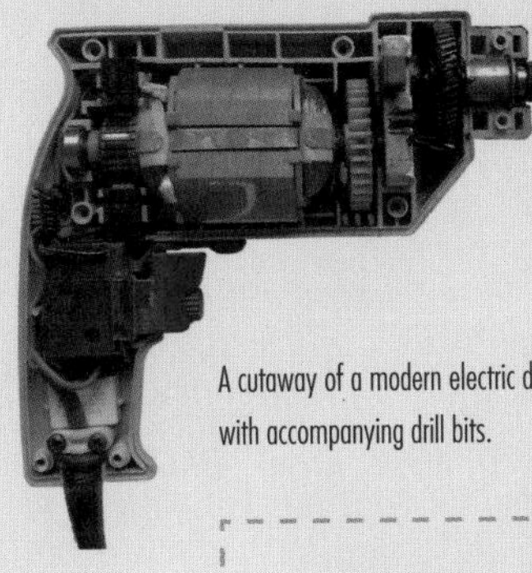

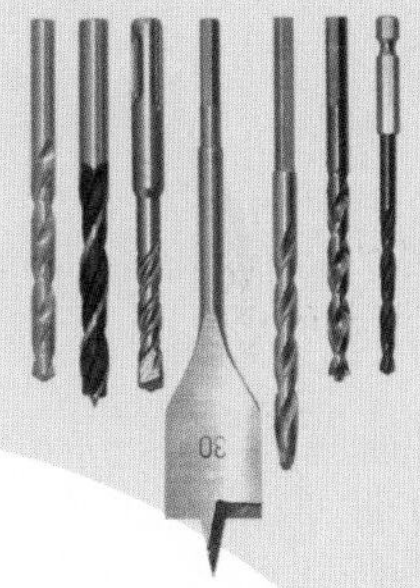

Although large and clunky to anyone familiar with today's streamlined plastic cordless drills, for its time, the B&D electric drill was a compact marvel. Decker designed the universal AC–DC motor that turned the chuck at 1,500 rpm. The drill bits were inserted into the chuck and secured with a chuck key as in the modern version. The only control was the trigger on/off switch, located on the inside of the "pistol grip," which allowed the operator to hold and operate the drill with one hand while using the other hand to hold the workpiece.

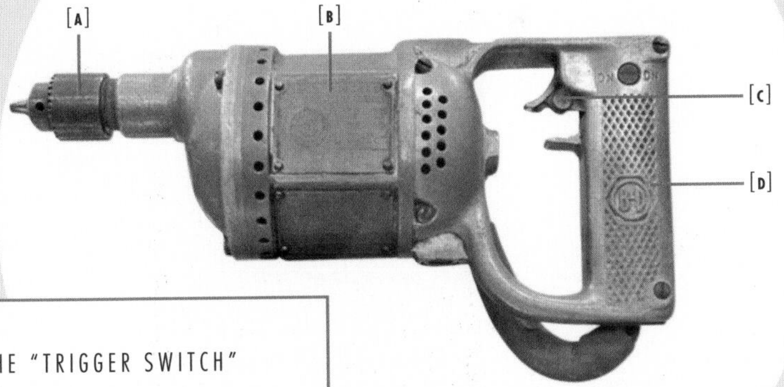

A cutaway of a modern electric drill, with accompanying drill bits.

[A]   [B]   [C]   [D]

KEY FEATURE: THE "TRIGGER SWITCH"

A single on-off switch is now standard on all machinery and domestic appliances, but until Decker patented his on/off trigger switch, electric equipment normally had both an on and an off switch. This simple innovation simplified the design of electrical equipment while simultaneously making it much more user friendly.

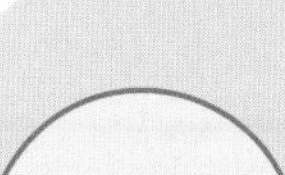

[A] Chuck
[B] Motor compartment
[C] Trigger on/off switch
[D] Grip

28

G.E. "MONITOR TOP" REFRIGERATOR

Manufacturer:
General Electric

Industry

Agriculture

Media

Transport

Science

Computing

Energy

Home ■

1927

By preventing food spoilage, the refrigerator has probably saved more lives than battalions of doctors and nurses. Although artificial refrigeration was widespread in the food and beverages industries by the late nineteenth century, it was only with the development of affordable, self-contained electric refrigerators in the twentieth century that the full benefits of refrigeration finally reached the home.

The iceman no longer cometh

Until the nineteenth century, cold drinks were freely available to all during the winter, at least in those latitudes where snow and ice were abundant, but in the warmer months, they were a luxury that only the wealthiest could afford. The idea of refrigeration using ice dates back to antiquity. Ice was collected in the winter and stored in icehouses or buried in deep pits to be used in the summer. With poor insulation, however, the wastage was enormous. With the vast increase in the urban population during the First Industrial Revolution, and the rise in living standards achieved by the Second, there was a steady growth in the demand for refrigeration: for the transport and storage of fresh food to the cities from distant production areas, and for luxuries such as cold drinks and ices.

REFRIGERATORS

Vapor-compression refrigeration	1834
Liquid-vapor refrigeration	1856
Gas-absorption refrigeration	1859
Linde ice machines	1876
G.E. Audiffren	1911
Kelvinator	1918
G.E. Monitor Top	1927

"It's always summertime in your kitchen. And the dangers of food contamination are always present—as long as it is possible for the temperature in your refrigerator to rise even a degree or two above fifty."

G.E. PRINT ADVERTISEMENT FOR THE MONITOR TOP REFRIGERATOR, 1929

G.E. "MONITOR TOP" REFRIGERATOR

Without any mechanical means to produce ice, however, the only option was to vastly increase the capacity of icehouses. The provision of ice to domestic customers became a huge business in the late nineteenth century. However, keeping a domestic "icebox" functional was labor intensive, and as demand continued to increase, there were problems in assuring sufficient supply. The development of mechanical refrigeration starting in 1834 eased the problem for business. The first industries to introduce refrigeration on a large scale were brewing in the 1870s and meatpacking a decade later. Industrial plants were much too large for home use, and the earliest domestic models had the cooling plant in the basement and the cold box in the kitchen. The first self-contained refrigerators appeared in the first decades of the twentieth century, with brands such as Frigidaire, Electrolux, and Kelvinator. However, with units costing more than a family car, the market for refrigerators remained tiny until G.E. introduced the first affordable model in 1927, nicknamed the "Monitor Top," which gradually reduced in price from $525 to $290.

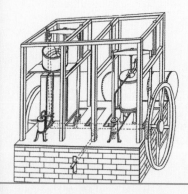

The precursor of the domestic refrigerator, John Gorie's ice machine of 1851.

THE G.E. "MONITOR TOP" REFRIGERATOR

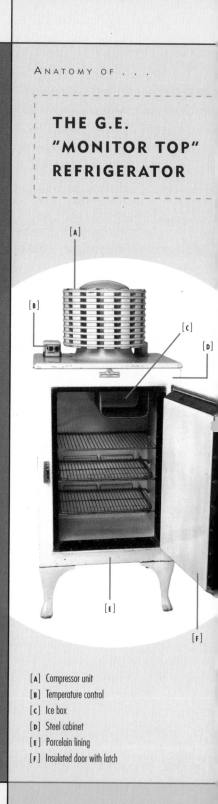

[A] Compressor unit
[B] Temperature control
[C] Ice box
[D] Steel cabinet
[E] Porcelain lining
[F] Insulated door with latch

The refrigerator takes advantage of the principle described by the Second Law of Thermodynamics that when a liquid changes to gas, it cools. Vapor compression remained the principle technology employed in domestic refrigeration used in domestic refrigerators for most of the twentieth century. Designed for General Electric by Christian Steenstrup (1873–1955), the Monitor Top was a self-contained refrigerator with a steel cabinet lined inside and out with white porcelain. A heavy door with a latch ensured that the unit was well insulated. Inside, the small icebox was just large enough to make a few trays of ice cubes, and three shelves provided storage space. The compressor and temperature control were on top of the unit. With its Chippendale legs, the Monitor Top looked more like a bedside or bathroom cabinet than a refrigerator, but in 1927 it was a cutting-edge design. While earlier models had been made of wood, the Monitor Top was made entirely of steel. The Monitor Top was produced in several sizes from 1927 until 1936, with one-, two- and three-door models. The unit used two toxic refrigerants, sulfur dioxide or methyl formate, which were replaced by non-toxic Freon in the 1930s.

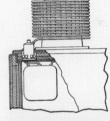

Schematic of the Monitor Top condenser on the top of the unit (left), and the condenser from a modern refrigerator that is fitted to the back of the unit (below).

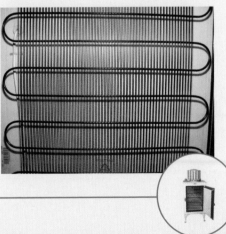

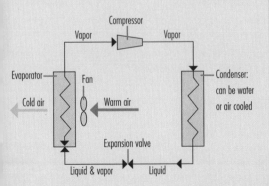

A schematic of the refrigeration cycle from a modern refrigerator.

G.E. "MONITOR TOP" REFRIGERATOR

29

LZ 127 GRAF ZEPPELIN

Manufacturer:

Luftschiffbau Zeppelin

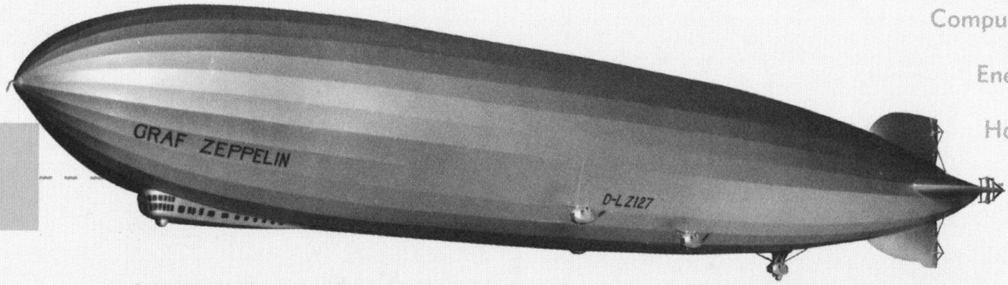

Industry

Agriculture

Media

Transport ■

Science

Computing

Energy

Home

"The 'Zeppelin spirit' grew out of the majesty, the incredible size, of the machine, along with its technological sophistication. Before World War I and afterward, it permeated the awareness of a German public eager to celebrate an engineering marvel that seemed 'all German.'"

G. DE SYON, *ZEPPELIN!* (2007)

1928

Ludwig Dürr

Until 1937, the wonders of the air were not fixed-wing heavier-than-air aircraft but the giant, stately liners of the skies, the zeppelins. Had they succeeded, surviving high-profile air disasters and the disinterest of the Nazis, zeppelins might have dramatically transformed the shape of air-passenger transport in the postwar world.

Magellan of the air

In 1929, the LZ 127 Graf Zeppelin, the most successful airship ever built in terms of flights achieved and passenger safety, completed the world's first circumnavigation of the globe by a lighter-than-air aircraft, with a flying time of 12 days, 12 hours and 13 minutes, and a total time for the journey of 21 days, 5 hours and 31 minutes, including stops at Friedrichshafen, Germany, Tokyo, Japan and Los Angeles,

California, before she returned to her starting point at Lakehurst, NJ. The achievement earned her captain and the head of Luftschiffbau Zeppelin, Hugo Eckener (1868–1954), the accolade of "Magellan of the air" and international recognition on both sides of the Atlantic. The trip, financed by U.S. press magnate William Randolf Hearst (1863–1951), ensured the survival of Luftschiffbau Zeppelin and its passenger services.

The zeppelins were built in giant sheds at their home port of Friedrichshafen, Germany.

AIRSHIPS

Montgolfière	1783
Steam-powered dirigible	1852
Dirigible No. 1	1898
LZ1 Zeppelin	1900
Baldwin dirigible	1908
LZ 127 Graf Zeppelin	1928

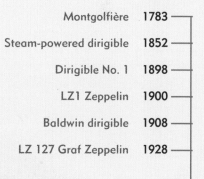

Although passenger airships are forever tainted by the Hindenburg disaster of 1937, we should remember that the early years of heavier-than-air aircraft also had their share of disastrous crashes. The curse of the airship, however, was its dependence on the highly inflammable gas hydrogen to provide its buoyancy. Although an alternative existed in helium, supplies of this safe inert gas were extremely limited, and in the 1930s the world reserves were held by the United States. However, the safety record of zeppelins alone cannot explain why the technology was abandoned after 1937. The LZ 127 had flown 9 years and completed thousands of passenger flights without major mishap. The true cause of the demise of the zeppelins was the rise of the Nazis, who came to power in Germany in 1933. They had little interest in developing airships, which they saw as too vulnerable to be used as weapons of war. And when the American government imposed an embargo on the export of helium to Germany in 1938, it was impossible to resume transatlantic flights between Germany and the U.S.

The passage of the zeppelins drew large crowds of spectators who marveled at the size and elegance of the giant aircraft.

THE LZ 127 GRAF ZEPPELIN

The gondola suspended under the main body of the zeppelin.

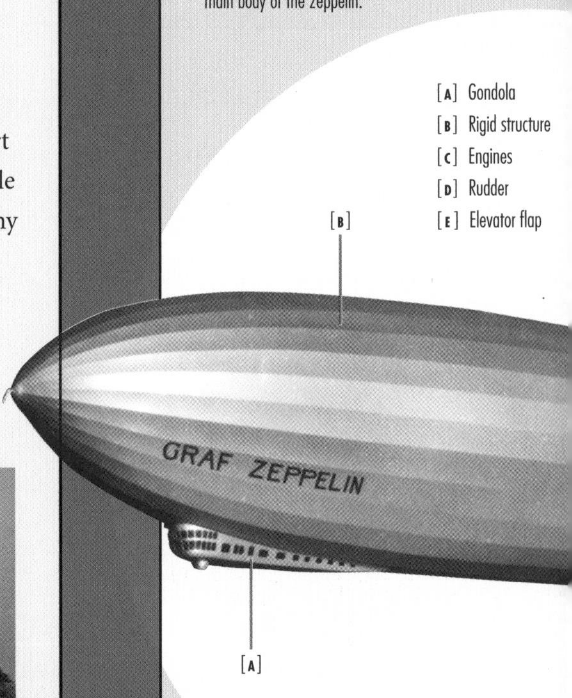

[A] Gondola
[B] Rigid structure
[C] Engines
[D] Rudder
[E] Elevator flap

[B]

GRAF ZEPPELIN

[A]

In his original conception of the airships that would bear his name, Count Ferdinand von Zeppelin (1838–1917) proposed linking several airships together into a "sky train." This intriguing idea was never tested; however, his rigid airship design, improved by Ludwig Dürr (1878–1956), was one of the most successful aircraft of the early twentieth century. At 776 ft (236.5 m), LZ 127 was as long as an ocean liner, but it had a fraction of a liner's capacity, with a crew of 40 and cabin space for 20 passengers, carried in the gondola suspended on the underside of the craft. In contrast to the cramped conditions of the propeller planes of the day, the passenger accommodation was luxurious, with two-berth cabins and a large panoramic saloon. Giant gas cells containing hydrogen and Blau gas accounted for the bulk of the enormous volume of the craft within the lightweight aluminum frame and cotton outer skin. Rigid airships were much stronger than their non- or semi-rigid predecessors, and therefore could be larger, faster, and lift heavier payloads. Powered by five 550-hp Maybach internal combustion engines burning Blau gas, LZ 127 had a top cruising speed of 80 mph (128 km/h).

KEY FEATURE:
BLAU GAS

While earlier zeppelins had used liquid fuel, this had the drawback that as the fuel was burned, the ship lost weight, requiring hydrogen to be continually vented. This problem was finally solved with the LZ 127, which used Blau gas as the fuel for its engines. Because the gas had the same density as air, it did not change the overall weight of the airship as it was used up.

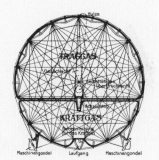

Cross-section of a zeppelin showing the different gas compartments.

[Right] Sections of the zeppelin's lightweight aluminum framework.

[Below] Graf Zeppelin was powered by five internal combustion engines.

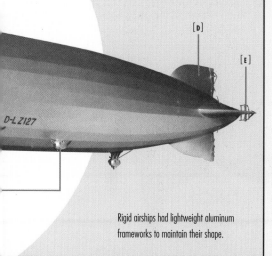

Rigid airships had lightweight aluminum frameworks to maintain their shape.

30

Designer:

John Logie Baird

BAIRD
"TELEVISOR"

Manufacturer:

Plessey Company

Industry

Agriculture

Media

Transport

Science

Computing

Energy

Home

1930

Television is without doubt the most influential media invention of the second half of the twentieth century. Written off as a fad that could never challenge the leading entertainments media of the day, radio and cinema, television had a difficult birth. Although it would quickly be superseded by electronic television, electromechanical television not only demonstrated the principle of the medium but also provided the platform to make and view the world's first television broadcasts.

Dramatic interludes

In the early years of broadcasting, the pioneers of television in both Britain and the United States decided to explore the medium's potential as a form of entertainment. The choice of material, however, was probably indicative of how the medium would develop in the two countries over the next few decades. The American offering, *The Queen's Messenger*, broadcast in 1928 from a studio in Schenectady, NY, was a thrills-and-spills action adventure drama. Two years later, the London-based British Broadcasting Corporation (BBC) opted for a highbrow one-act play about death, *The Man with the Flower in His Mouth*, by Nobel laureate Luigi Pirandello (1867–1936). However, because of the severe technical limitations of electromechanical television, putting on any kind of dramatic programming was extraordinarily ambitious.

The groundbreaking American broadcast used two actors and three cameras, but the screen was so small that viewers watching on their primitive sets could only see either a single actor's face or his hands. The director used two of the cameras to film the actors' faces, while the third moved between their hands and the selection of props required by the script. A large crew of technicians provided the sound and visual effects. In its first dramatic broadcast, the BBC used a single Baird camera that provided the only light source in an otherwise darkened studio to film the cast of three, who played live in front of painted black-and-white backdrops representing a sidewalk café. In both cases the sound and image were broadcast separately and synchronized in the television receiver.

The *New York Herald Tribune*'s verdict on the world's first television play was lukewarm. It concluded: "It was the general opinion among those that watched the experiment that the day of radio moving pictures was still a long, long way in the future. Whether the present system can be brought to commercial practicability and public usefulness remains a question."

BAIRD "TELEVISOR"

Round and round she goes

Readers may wonder why I picked an electromechanical device to represent television, instead of one of the electronic models that dominated the postwar market until the advent of digital TVs. I defend my choice because all the television firsts, including interviews, sports, music, drama, outside broadcasts and coast-to-coast and transatlantic transmissions, as well as color and video recording, were all achieved with electromechanical television, broadcasting mostly in monochrome, and with an extremely low image resolution of between 30 and 240 lines, in the case of the Baird system. However, even with these technical limitations, by 1936, electromechanical television had firmly established the medium in the popular imagination. Subsequent developments merely built on what Baird and other TV pioneers had achieved.

The first TV image broadcast by John Logie Baird.

TELEVISION

- **1884**   Nipkow disc
- **1906**   Rosing mechanical TV
- **1907**   Electronic TV
- **1922**   Iconoscope
- **1925**   Jenkins mechanical TV
- **1925**   Color TV tube
- **1926**   Baird mechanical TV
- **1926**   Radioskop
- **1927**   Image Dissector
- **1930**   Baird Televisor

The Baird Televisor and other similar devices are called electromechanical televisions because they combined moving mechanical parts with electronic components. The heart of the system was the Nipkow disc, named for its German inventor Paul Nipkow (1860–1940). The spinning perforated disc acted as a basic scanner, breaking down an image so that it could be transmitted electronically to a receiver equipped with another Nipkow disc. Nipkow came up with the idea when he was a student in 1883 but he never developed the idea, which was picked up decades later by the Scottish inventor John Logie Baird (1888–1946) in Britain and by Charles Jenkins (1867–1934) in the U.S. Nipkow lived long enough to see the full realization of his invention, when in 1928, aged 68, he saw a demonstration of the Baird system in Berlin.

Three years after a successful demonstration of synchronized sound and television pictures in 1925, Jenkins set up the first television broadcasting station in the U.S., transmitting on five evenings a week. However, by 1932, his company had gone bust, and Jenkins himself died 2 years later. Baird, with the backing of the BBC, fared much better, and his electromechanical TV continued to broadcast until 1937.

> "Television won't be able to hold on to any market it captures after the first six months. People will soon get tired of staring at a plywood box every night."
>
> DARRYL F. ZANUCK IN AN INTERVIEW IN 1946

Solid state

Although electromechanical television was an important breakthrough, there was one thing missing to confirm its significance: patent infringement disputes. As far as I am aware, there was no litigation between Baird and Jenkins despite the similarity of their systems. The major players in television were already looking beyond electromechanical systems to electronic television. Just as Jenkins and Baird were making their first tentative broadcasts, the race to perfect electronic television was entering its final stages. The three major players were Philo Farnsworth (1906–1971) and Vladimir Zworykin (1888–1982) in the U.S. and Kálmán Tihanyi (1897–1947) in Hungary. Farnsworth was the first to demonstrate electronic television with his "Image Dissector" between 1927 and 1929, while Zworykin was working on the "Iconoscope" while working for Westinghouse. Although Zworykin could claim

a prior demonstration and patent in 1925, his image had been static. The conflicting claims led to protracted patent litigation between Farnsworth and RCA, who had acquired Zworykin's patent, which was not settled in Farnsworth's favor until 1939. But it was only when the Farnsworth and Zworykin patents were combined with Tihanyi's revolutionary camera tube, developed for his "Radioskop" system in 1926, that the development of electronic television was finally complete.

Zworykin demonstrates the successor of electromechanical television, the electronic cathode ray tube TV.

BAIRD "TELEVISOR"

ANATOMY OF . . .

THE BAIRD TELEVISOR

[A] Screen
[B] Nipkow disc
[C] Electric motor
[D] Neon lamp
[E] Casing for Nipkow disk
[F] On/Off Switch
[G] Radio Receiver/Tuner

[G]

[F]

[E]

[A]

Front view,
case removed

Rear view,
case removed

[B]

[C]

[D]

The Baird Televisor was initially sold in kit form with all parts exposed. The first production model released in 1930 encloses the working parts in a metal cabinet, which looks like a strange hybrid between an old-style kitchen stove and a penny-in-the-slot arcade peep show. The circular section at the center housed the nipkow disc, powered by an electric motor. The "screen" to the right of the Televisor consisted of a magnifying lens in front of the disc and a neon lamp behind it. The screen image was made up of 30 lines (although the Baird system increased its resolution to 240), displayed vertically rather than horizontally as in today's TVs. The small portrait display was only large enough for one person to see at a time. Instead of being black and white, the image was orange and white because of the color of the neon tube. The image was transmitted by radio waves and a separate set was used to receive the sound for the transmission. Although about 1,000 Televisors were manufactured, Baird himself knew that it was a crude prototype.

KEY FEATURE: THE NIPKOW DISC

The Nipkow disc both created and reproduced the image.
In the studio, light was projected through the disc onto the subject.
Photoelectric cells converted the variations in reflected light into
pulses that were amplified and transmitted by radio waves.
The pulses caused the neon light in the Televisor to flash behind
the disc, recreating the original image on the screen.

Nipkow disc

John Logie Baird working on an electromechanical television. In this experiment, he is scanning his hand with the Nipkow disc and broadcasting the image by radio.

31

Designer:
Willis Carrier

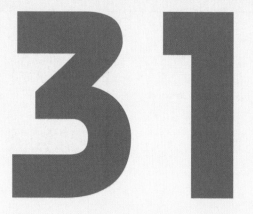

PHILCO-YORK "COOL WAVE" AIR CONDITIONER

Industry
Agriculture
Media
Transport
Science
Computing
Energy
Home

Manufacturer:
York Ice Machinery Company

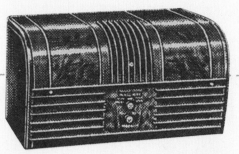

"It was luxuries like air conditioning that brought down the Roman Empire. With air conditioning their windows were shut, they couldn't hear the barbarians coming."

GARRISON KEILLOR (B. 1942)

1938

Air conditioning has important industrial applications in fields as diverse as food production and computer hardware manufacture, but for the public, its main impact has been to improve the quality of life in the warmer parts of the world, notably in the U.S.'s "Sun Belt," where the introduction of air-conditioning was a necessary pre-condition to large-scale population migration and economic development.

Willis Carrier

The southern wave

Raised in one of the cooler parts of the world, where heat waves last days and not weeks or months, I did not grow up appreciating the benefits of central air. However, when my parents moved to central Texas, where the average summer temperature can reach as high as 87°F (31°C), the attractions of air condition-ing immediately became apparent. Although domestic refrigeration became common in the 1920s, domestic air-conditioning lagged behind by a decade, with the first portable window air-conditioning unit, the Philco-York "Cool Wave," going on sale in 1938. The Cool Wave was a joint venture between the Philco Company, which was a major U.S. radio manufacturer, and the York Ice Machinery Company, which specialized in industrial refrigeration and air-conditioning systems.

Although I have listed Willis Carrier (1876–1950) as the designer, specialists in the history of air conditioning will know that he did not design the Cool Wave (whose designer remains sadly anonymous), but was the inventor of air conditioning itself, and the founder of a rival to the York Ice Machinery Company, the Carrier Company. Carrier patented his first industrial air-conditioning system in 1906, which he had designed for a printing company in Buffalo, New York, in 1902. An air conditioner works much like a refrigerator, with a liquid coolant circulating through coils under pressure. When the coolant passes through an expansion valve, suddenly reducing its pressure, the tempera-ture drops, cooling the air that is blown in the room by a fan.

In 1930, Thomas Midgley (1889–1944) developed the CFC Freon to replace toxic coolants, such as ammonia, previously used in refrigerators and air conditioners. The Cool Wave combined Carrier's basic design with Midgley's Freon and packaged it in a stylish wooden cabinet. The unit could be plugged into an electrical socket and was small enough to move from room to room. The era of home "summer comfort" had arrived, and with it, the opening of America's "Sun Belt" to immigration from the northern states and economic development.

PHILCO-YORK "COOL WAVE" AIR CONDITIONER

32

Designer:
Ernst Ruska

SIEMENS
ELECTRON
MICROSCOPE

Manufacturer:

Siemens-Reiniger-Veifa MbH

Industry
Agriculture
Media
Transport
Science ■
Computing
Energy
Home

1939

The development of the electron microscope overcame the limitations of optical microscopy using light. The transmission electron microscope enabled major advances in medical science with the imaging of the structure of the cell and viruses, and in engineering and physics, by revealing the atomic structure of materials.

A visible need

In the summer of 1930, the German electrical engineer Reinhold Rudenberg (1883–1961) received the fateful news that his infant son had contracted polio. Until 1950, when an effective vaccine was developed, polio was often fatal or left its victims with paralysis and wastage of the legs. Although the polio virus was known, it and other viruses were too small to be visible to the optical microscopes of the day. Rudenberg, building on existing theoretical research, proposed a microscope that would replace light with a beam of electrons that could be focused by electrostatic lenses. His employer, Siemens AG of Germany, took out patents on the principle in 1931.

The first optical microscopes are thought to have evolved from early telescopes in the late sixteenth century. But their full impact was only realized in the seventeenth century with the work of Robert Hooke (1635–1703) and of the man described as the "father of microbiology," the Dutchman Antonie van Leeuwenhoek (1632–1723), whose publications were among

the first to reveal the microscopic world to the general public. The magnifying power of a compound optical microscope is around x1,000, therefore insufficient to reveal biological structures as small as viruses. In 1897, the physicist J. J. Thomson (1856–1940) discovered the electron, and in the following decades, several scientists, including the Hungarian physicist Leó Szilárd (1898–1964), proposed using this subatomic particle, which has a much shorter wavelength than visible light, as a means of magnification.

In 1931, while he was a postgraduate student, Ernst Ruska (1906–1988) built a prototype transmission electron microscope (TEM). He continued to develop the TEM during the 1930s and designed the first commercial device in 1939 by taking advantage of the Siemens patents after he joined the firm in 1937. Realizing Rudenberg's dream, Ruska, working with his brother, the microbiologist and doctor Helmut Ruska (1908–1973), was able to produce the first images of viruses.

SIEMENS ELECTRON MICROSCOPE

[A] Cathode
[B] Gas discharge tube
[C] Condenser lens
[D] Objective lens
[E] Projection lens
[F] Glass plate
[G] Object plane
[H] Plane of intermediate picture
[I] Observation window

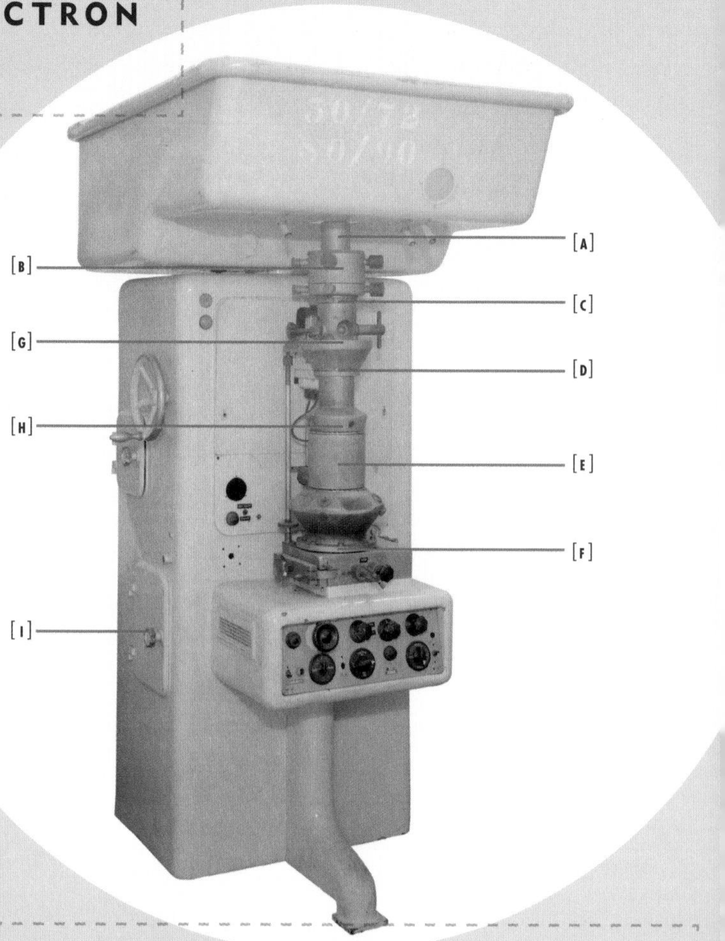

[B]
[G]
[H]
[I]

[A]
[C]
[D]
[E]
[F]

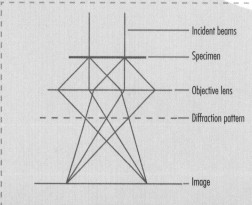

Incident beams

Specimen

Objective lens

Diffraction pattern

Image

This sketch shows how the parallel beams of electrons pass through the specimen and are diffracted in various directions. The objective lens collects the beams originating from the same point of the sample and focuses them onto the image plane. Observing the electrons in this plane reveals the diffraction pattern of the electrons.

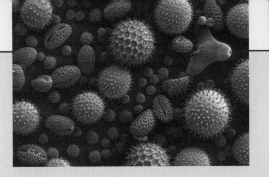

A transmission electron microscope consists of three main sections: the illuminating apparatus, which is a high-voltage electron gun (cathode) located at the top of the unit; the specimen stage; and the lens assembly, consisting of electromagnetic lenses that form and focus the electron beam. In the TEM, the sample is an extremely thin slice of material, part of which is transparent to the electron beam. The beam will pass through the specimen, collecting structural information about the material, which is magnified by the electromagnetic lenses. The data is transmitted to the recording/viewing system. The image thus produced can be displayed on a fluorescent viewing screen or recorded photographically (though in modern designs, it is transmitted directly to a computer display). Ruska's first prototype had a magnification lower than the best

Pollen grains magnified 500 times by electron microscopy.

optical microscopes of the day, but having established the principle of electron microscopy, in the next few decades, after the war he developed the TEM to give a magnification of x100,000. The much higher magnifications possible with today's TEMs have not only transformed medicine but also materials science, as the instrument can be used to study the structure of materials at the atomic level.

"Regarded by some scientists as the most important invention of the 20th century, the modern electron microscope allows magnifications of up to 2 million times. While other inventions have had much greater social and cultural impacts, the electron microscope has become a crucial tool in numerous scientific fields." R. CARLISLE, *SCIENTIFIC AMERICAN INVENTIONS AND DISCOVERIES* (2004)

KEY FEATURE:
THE ELECTROMAGNETIC LENSES

A magnetic lens is designed to focus the electron beam like a glass lens focuses light. The TEM has three types of lens: condenser lenses that form the electron beam; objective lenses that focus the beam that comes through the sample; and projector lenses that transmit the image onto the screen or film. The lenses consist of electromagnetic coils arranged in a square or hexagonal configuration.

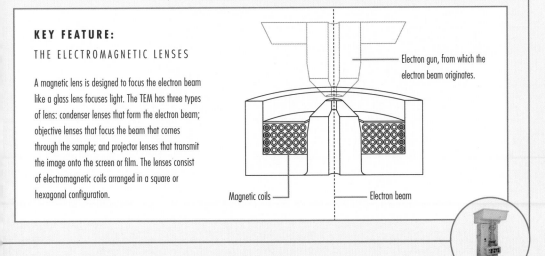

Electron gun, from which the electron beam originates.

Magnetic coils

Electron beam

33

Designer:
Wernher von ßraun

MITTELWERK
V-2 ROCKET

Manufacturer:

Mittelwerk GmbH

Industry

Agriculture

Media

Transport

Science ■

Computing

Energy

Home

1944

Although initially developed as a terror weapon by the Nazi regime during the Second World War, the V-2 rocket became the basis for the space programs of the U.S. and the USSR. At the end of the war, V-2 designer Wernher von Braun and members of his team surrendered to the Americans, giving the U.S. a significant advantage in rocket technology.

Light the fuse and run

A rocket, for all intents and purposes, is a hollow metal tube filled with a highly explosive substance, which, when ignited propels the rocket skyward at tremendous speed. The trick is, of course, to prevent the rocket from exploding (a) on the ground, or (b) in flight before it reaches its destination/target, or (c), in the case of a manned space vehicle, at all. The Chinese, who invented gunpowder, were the first to experiment with rockets as weapons around the twelfth century. They solved problems (a) and (b), but a solution to (c) would have to wait until the mid-twentieth century and the development of the Russian and American space programs.

The man hailed as the father of modern rocketry is the American physicist Robert Goddard (1882–1945) who designed the first liquid-fuelled rockets in the 1920s. One man who was particularly interested in Goddard's work was a young German rocket engineer, Wernher von Braun (1912–1977). In 1933,

von Braun became a member of the Nazi Party and of the paramilitary SS, though he claimed later that he had joined these organizations to be able to continue his rocket research and did not take part in any political activities. The V-2, however, was built with slave labor, and is one of the few weapons whose construction killed more people (est. 20,000) than the fatalities it caused in action (est. 7,250). Once working for the U.S., however, von Braun was exonerated of any wrongdoing during the war.

Although the V-2 was the most advanced rocket of its time, the damage it caused to the Allied war effort was so negligible that its deployment in the fall of 1944 probably shortened the war by taking away resources from the construction of tactically more valuable fighter aircraft. After Germany's defeat, the Americans and Russians both acquired V-2 technology, which later became the basis of their space programs. The first successful test flight of a V-2 in 1942 is taken to mark the beginning of the Space Age.

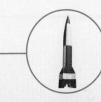

THE MITTELWERK V-2 ROCKET

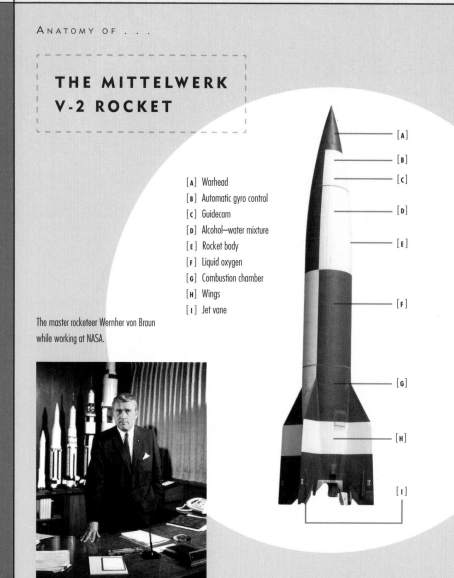

[A] Warhead
[B] Automatic gyro control
[C] Guidecam
[D] Alcohol–water mixture
[E] Rocket body
[F] Liquid oxygen
[G] Combustion chamber
[H] Wings
[I] Jet vane

[A]
[B]
[C]
[D]
[E]
[F]
[G]
[H]
[I]

The master rocketeer Wernher von Braun while working at NASA.

"We knew that each V-2 cost as much to produce as a high-performance fighter airplane [....] From our point of view, the V-2 program was almost as good as if Hitler had adopted a policy of unilateral disarmament."

DISTURBING THE UNIVERSE (1979) BY F. DYSON

Like many modern missile systems, the V-2 was a mobile weapon.

The V-2 was just shy of 46 ft (14 m) long and weighed 13.7 tons (12.5 metric tons). When fitted with a warhead, it was the first long-range ballistic missile, capable of delivering a payload of 2,200 lb (1,000 kg) of conventional explosives, with a maximum range of 200 miles (320 km), traveling at 3,580 mph (5,760 km/h). In order to escape detection by enemy aircraft, the V-2 was launched from a mobile launch platform, the Meillerwagen. Had the Germans succeeded in developing the A-bomb, they could have used the V-2 to obliterate London and Moscow, and even attack the continental U.S., after a successful launch from a platform towed by a U-boat. Fortunately for the world, the Germans, like the Iranians today, built a missile before developing its atomic payload. When launched from the Meillerwagen, the V-2, guided by gyroscopes, had a range of 55 miles (88 km), but if launched vertically, it could reach an altitude of 128 miles (206 km), which is more than twice the 62-mile (100-km) boundary of Earth's atmosphere. On October 24, 1946, a camera mounted on an American V-2 took the first pictures of Earth from space.

KEY FEATURE:
THE LIQUID-FUEL ROCKET ENGINE

The V-2 engine was fueled by 8,400 lb (3,810 kg) of ethanol/water and 10,800 lb (4,910 kg) of liquid oxygen, giving it a total burn time of 65 seconds. Hydrogen peroxide steam turbines pumped the fuel and oxygen into the combustion chamber, and then into the burner chamber through 1,224 nozzles that ensured the correct mix of ethanol and oxygen.

A V-2 rocket engine on display at the National Museum of the United States Air Force in Dayton, Ohio.

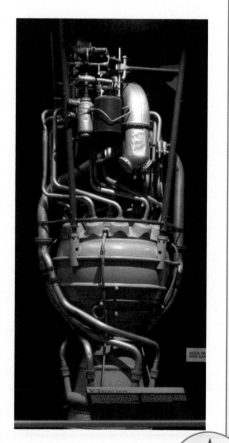

34

Designer:
General Electric
R&D Dept

G.E. AUTOMATIC TOP-LOADING WASHER

Manufacturer:
General Electric

Industry
Agriculture
Media
Transport
Science
Computing
Energy
Home

1947

Ten years after giving the homemaker the first affordable, self-contained refrigerator, the "Monitor Top," General Electric developed the most important labor-saving appliance of the century: the automatic top-loading washer, disposing at a stroke of one of the week's most time-consuming and irksome tasks.

Taking a load off

If you'd asked your great grandmother what the worst household chore was, she probably would have replied, the weekly wash. Although the first electric washing machines had been introduced at the beginning of the twentieth century, early models were little more than primitive hot tubs with moving paddles with few of the functions we take for granted on washers today. They agitated the clothes in hot soapy water, but they did not fill or empty automatically, which had to be done by hand, or spin the wash to remove excess water. By the 1940s, washers were fitted with electric wringers (mangles) to wring out clothes before they were hung out to dry. Combo washer-dryers were still a decade or so away. What takes a modern automated front or top loader 30–45 minutes, depending on the cycle chosen, could take a homemaker two hours or more with a pre-1947 washer.

"Without wetting a hand, the user of an automatic can do a nine-pound wash in half an hour, a chore that still takes two hours using the conventional washers, according to time and motion studies."

"How to Choose a Washer" from Popular Science (1947)

An early nineteenth-century manual drum washer—the ancestor of the modern washer.

The prehistoric human-powered washer: the Old Alfa

As we've seen in earlier entries on the sewing machine, the typewriter, the vacuum cleaner, and the refrigerator, women's roles had begun to change in the late nineteenth century. Women's emancipation continued after the First World War, and accelerated after the Second: Women were much more likely to go to work and much less likely to have help at home, though they were still expected to do the bulk of the cooking, cleaning, and childcare. By the mid-1940s, improvements in technology combined with increasing disposable incomes meant that labor-saving domestic appliances were now within reach of most middle-class families. The Holy Grail of laundry in the postwar U.S., which was finally delivered by the G.E. Automatic Washer in 1947, was a fully automated cycle that you turned on and came back to when the clothes had been washed, rinsed, and spun dry, ready to be aired or ironed.

ANATOMY OF . . .

THE G.E. AUTOMATIC WASHER

Close-up of the temperature and timer controls.

KEY FEATURE:
AUTOMATION

The individual components of the Automatic Washer were not new, but what was novel was their assembly into a fully automated mechanism that allowed the homemaker to start the machine and return 45 minutes later to clean and near-dry clothes.

The tub with central agitator and internal soap dispenser.

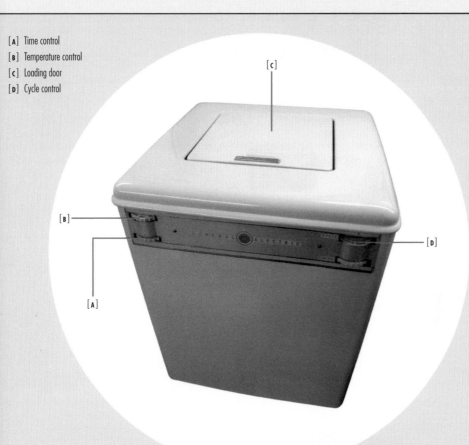

[A] Time control
[B] Temperature control
[C] Loading door
[D] Cycle control

In terms of design and functions, the G.E. Automatic Washer (AW) marked the beginning of modern laundry appliances. The machine had two sets of controls: on the left, the temperature selector, with a choice of "warm," "medium" and "hot," and the wash timer, ranging from 3 to 20 minutes. The cycle selector on the right featured a choice of full automatic cycle (45 minutes), or manual settings for "soak," "wash," "rinse," and "dry." Once the clothes were loaded, detergent was added to the internal soap dispenser. The AW was plumbed into the house's water supply and drainage, therefore did not require manual filling and emptying. The water entered via the soap dispenser into the machine. When it reached the correct level, the water flowed into the top of the agitator and down into a cup that started the washing action. At the end of the wash the AW drained and rinsed the inner tub of suds before initiating a first spin. After completing the rinse cycle, the machine spun at 1,140 rpm, drying the clothes in the inner tub. Water remained in the outer tub and could be either drained by selecting "empty" on the main control, or reused for another cycle.

35

Designer:

Jack Mullin

AMPEX MODEL 200A
TAPE RECORDER

Manufacturer:

Ampex Electric and Manufacturing Company

Industry

Agriculture

Media ■

Transport

Science

Computing

Energy

Home

1948

Until the end of the Second World War, sound recording and broadcasting was limited by its dependence on record discs, which were difficult to edit and had poor sound quality. The introduction of reel-to-reel tape-recorder technology in 1947 simplified sound editing, improved broadcast quality, and made possible the wide use of prerecording for radio broadcasts.

Adolf and Bing

Would-be world conqueror and crazed homicidal dictator Adolf Hitler (1889–1945) and singer, actor, and entertainer Bing Crosby (1903–1977) are not two names that are usually closely associated. However, both contributed in their own way to the postwar revolution in sound recording and broadcasting. During the closing years of the Second World War, Jack Mullin was a young officer in the U.S. Army Signal Corps stationed in England working to prepare for the D-Day landings. As he worked late into the night, he would listen to high-quality German radio music broadcasts that he realized were far better than any "canned" music broadcast at the time in the U.S. or UK. When Germany was defeated, he was sent to France and Germany to investigate the German military's top-secret electronics equipment. It was by chance, when he visited a radio station near Frankfurt, that he stumbled across a German Magnetophon, a high-fidelity reel-to-reel tape recorder using an early version of magnetic tape. Realizing the machine's potential, Mullin acquired two for the U.S. government, and another two for himself, which he dismantled and shipped back to San Francisco.

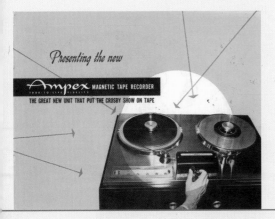

The push-button controls now seen on all media devices were revolutionary in 1948.

Having rebuilt and improved his two Magnetophons, Mullin began to demonstrate them in 1946. In 1947, he gave a demonstration to Bing Crosby, then America's most popular radio and movie star. Crosby, who did not like the pressure of live broadcasting that the networks insisted on because disc recordings were so poor, had temporarily retired from live radio work. Impressed by Mullin's tape recorder, Crosby hired him to record and edit his shows for the 1947–1948 season. He later invested $50,000 to develop Mullin's prototypes into the first American-made reel-to-reel tape recorder, the Ampex Model 200/200A. Mullin kept the first two Model 200s and a further 12 went in service at ABC's studios in 1948.

THE AMPEX MODEL 200A

KEY FEATURE:
MAGNETIC FERRIC OXIDE TAPE

According to Mullin the most important feature of the new machine was magnetic tape, which gave to artists and broadcasters the freedom to prerecord and edit material before it was broadcast. He described how he himself invented tape-editing techniques by trial and error while recording the first season of the *Bing Crosby Show* for ABC. On one occasion, he spliced in laughter into the recording of a show that had not got many laughs, creating the first "laugh-track." When he first demonstrated the Magnetophon, Mullin used the wartime BASF ferric oxide tape, but from 1948 the Model 200A used the American-made 3M Scotch 111 gamma ferric oxide-coated acetate tape.

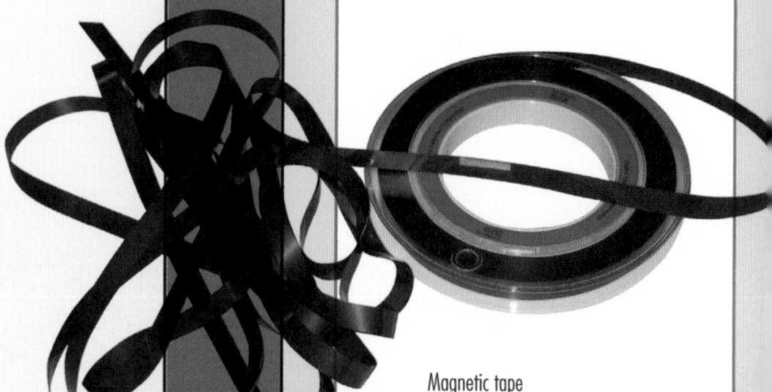

TAPE RECORDERS

- 1886   Wax-strip recorder
- 1898   Telegraphone
- 1930   Blattnerphone
- 1935   Magnetophon
- 1948   Ampex 200A

Magnetic tape

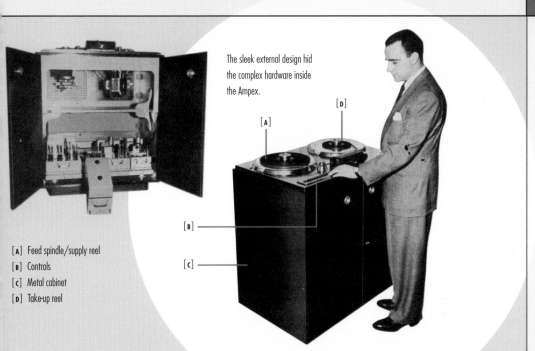

The sleek external design hid the complex hardware inside the Ampex.

[A] Feed spindle/supply reel
[B] Controls
[C] Metal cabinet
[D] Take-up reel

"In Germany […], Hitler could have anything he wanted. If he wanted a full symphony orchestra to play all night long, he could get it. Still, it didn't seem very likely that even a madman would insist on live concerts night after night. There had to be another answer, and I was curious to know what it was." **J. T. MULLIN**

Although described as a "portable," a better term for the Ampex Model 200A would be "mobile." Mounted on a metal cabinet, the Model 200A had all the features that would be found on truly portable reel-to-reel machines (made by Philips in 1951) and on cassette tape recorders. Unlike the complicated looking Magnetophon on which it was modeled, the Model 200A had a sleek, pared-down design. A single control panel on the side of the tape recorder featured five illuminated transparent pushbuttons: "start," "stop," "rewind," "fast forward" and "record." The 35-minute ¼-in (6.3-mm) tape was mounted on the feed spindle, threaded around the idler, past the three tape heads (erase, record, play) and the capstan and was wound around the take-up reel. The machine had an automatic cut-off at the end of recording, as well as an optional double-speed rewind function.

The Second World War German AEG Magnetophon.

36

Designer:
Ronald Bishop

DE HAVILLAND
DH106 COMET

Manufacturer:

De Havilland

Industry

Agriculture

Media

Transport

Science

Computing

Energy

Home

1949

Developed during the Second World War as secret weapons, jet aircraft really came into their own after the war. The first passenger jetliner, the de Havilland DH106 Comet, revolutionized air travel when it went into service in 1952. However, its innovative design was fatally flawed, leading to several high-profile crashes that allowed the de Havilland's American competitors to take the lead.

Technical highs and lows

The British are noted for their technological innovations. During the First Industrial Revolution, Britain led the world in science and technology, and as a result was the world's economic, military, and political superpower for most of the nineteenth century. From the mid-twentieth century, however, though British engineers carried on developing world-beating inventions, British business failed to capitalize on them, letting overseas competitors steal the prize, the cash, and the glory.

A case in point was the development of the first commercial jetliner, the DH106 Comet, which came into service with BOAC, the forerunner of today's British Airways, in 1952. Despite the initial success of the plane, the entire Comet fleet had to be grounded 2 years later after a series of disastrous accidents. The Comet crashes did not lead to the suspension or scrapping of passenger jet transport in the same way as the Hindenberg disaster of 1937 heralded the end of the era of passenger airships. But it allowed Britain's competitors, notably Boeing, McDonnell Douglas, and Lockheed, to seize the technological and commercial initiative and to dominate the jetliner market for the next five decades. The Comet was an object lesson in how great technological innovation does not necessarily lead to commercial success.

JET AIRCRAFT

Heinkel He 178	1939
Caproni Campini N1	1940
Gloster Whittle	1941
Messerschmitt Me 262	1942
Gloster Meteor	1943
Lockheed P-80	1944
De Havilland Vampire	1945
Viking VC1	1948
DH106 Comet	1949

DE HAVILLAND DH106 COMET

The Comet, designed by Ronald Bishop (1903–1989), was a bold technological gamble that used a new and relatively untested propulsion technology, the jet engine. The first jet aircraft had only been developed less than a decade earlier, during the Second World War, when both the Allies and the Axis powers had produced jet fighters and bombers. Although the Germans were the first to put jet fighters into service, they came too late in the war to save the Nazis from defeat. The British government's decision to back the project for the first civilian jetliner was not only unusually far-sighted but also extremely courageous.

The Comet cockpit may look primitive but was state of the art in 1949.

Crash and burn

Although the Comet was the most rigorously tested plane yet built, within 6 months of entering into service, a BOAC flight departing Rome airport failed to take off and overshot the runway, injuring two passengers. It was the first of many accidents that plagued the aircraft's first 2 years in service. Several accidents were due to human error by flight crews more used to flying propeller aircraft, and several others to poor weather. However, between January 1953 and April 1954, three Comets disintegrated in flight, killing everyone on board. The entire fleet was grounded and the salvaged wreckage was subjected to minute examination. At the time, the grounding of the Comet was seen as a national disaster. The British premier Sir Winston Churchill (1874–1965) wrote: "The cost of solving the Comet mystery must be reckoned neither in money nor in manpower."

The large square panoramic windows catastrophically weakened the fuselage.

"The most brilliant [...] de Havilland effort in the postwar years was undoubtedly the stunning DH106 Comet [....] Although an otherwise masterful design, it fell prey to lack of experience in building large airliners with pressurized cabins and encountered fatigue problems that caused crashes."

AIR WARFARE (2002) BY W. BOYNE

The court of inquiry concluded that the crashes had been caused by excessive stress on the fuselage, especially around one of the signature features of the aircraft: the large square windows. The first generation of Comets was taken out of service and scrapped, and the aircraft was completely redesigned. A new, larger Comet 4 was produced for both military and civilian use, starting in 1958. The last passenger model was taken out of service in 1997, and the last military model operated by the Royal Air Force, in 2011.

Although the Comet returned to service with a strengthened fuselage and smaller round windows, the damage to the plane and Britain's reputation for technological excellence had been done. The British aviation industry would never again challenge its American competitors. Among the most enduring legacies of the Comet tragedy are the under-wing nacelle engine configuration seen on all large commercial airliners and small porthole windows.

DE HAVILLAND DH106 COMET

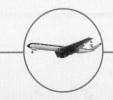

> "The cost of solving the Comet mystery must be reckoned neither in money nor in manpower."

WINSTON CHURCHILL, (1874–1965)

Traveling in style

If you saw a Comet taxiing on an airport runway today, it would not look out of place among modern passenger jetliners—the reason being that it set the standard for aircraft design for decades to come and was far in advance of any of its propeller-driven predecessors and rivals. Once aboard, however, a passenger in one of today's overcrowded medium-range aircraft, with their cramped seating, narrow aisles, and shaky foldaway tables, would envy the internal layout and facilities of the Comet.

Although the Comet is roughly the same size as the Boeing 737 and the Airbus 320, these planes were designed for the mass-air transit age to carry more than 100 passengers. The Comet's initial configuration specified 11 rows of two pairs of seats divided by a broad aisle. BOAC and Air France, however, decided on an even more spacious cabin layout with 36 seats. The fully pressurized cabin was much quieter than any prop plane of the time. The on-board facilities included a galley kitchen with meals served on china and eaten with silverware, separate men's and women's bathrooms, and the fatal square panoramic windows. Its safety features included life rafts stored in the wings and life vests stowed under each seat.

ANATOMY OF . . .

THE DH106 COMET

KEY FEATURE:
THE GHOST MK 1 JET ENGINES

A feature not seen in later large passenger jetliners is the mounting of the four engines—two pairs of de Havilland Ghost 50 Mk1 turbojets—inside the wings abutting the fuselage. The designers chose this configuration because it greatly reduced the drag when compared to engines suspended underneath the wings or on the fuselage, making the plane faster and more fuel-efficient. The position of the engines also reduced the risk of foreign object strikes—a major problem for turbine engines—and it also facilitated engine maintenance; however, it increased the risk of a catastrophic failure if an engine caught fire or exploded in flight.

Although very familiar to the twenty-first-century airline passenger, the Comet's streamlined design would have appeared revolutionary and futuristic in 1952. The plane had a spacious tubular fuselage 94 ft (29 m) long, fronted by a conical cockpit for a crew of four, v-shaped swept-back wings, with a wingspan of 115 ft (35 m), a tail fin and rudder, and the horizontal stabilizer mounted on the fuselage. The passenger entry door was at the rear of the aircraft. Everything had been done to minimize the weight and drag of the aircraft. The outer skin was made of a new lightweight aluminum alloy. The four engines were mounted internally, and the fuel tanks with a capacity of 27,300 liters (7,212 U.S. gallons) were inside the wings.

The four engines gave the Comet a top speed of 450 mph (724 km/h), which halved the flying time across the Atlantic, although because of the plane's limited range of 1,500 miles (2,414 km), this advantage was offset by the need for several refueling stops.

[A] Cockpit
[B] Passenger compartment (36–44 seats)
[C] Square windows
[D] Wing and fuel tanks
[E] Ghost Mark 1 engines
[F] Passenger door
[G] Tail fin/rudder
[H] Horizontal stabilizer

37

Designer:
Mervyn Richardson

VICTA ROTOMO
FAN MOWER

Manufacturer:
Victa Mowers Pty. Ltd.

Industry

Agriculture

Media

Transport

Science

Computing

Energy

Home

1954

In previous entries we have seen several labor-saving appliances aimed at women, but with the Victa Rotomo, we feature an Australian invention that would improve the lot of husbands and fathers the world over. If women could be freed from the washday blues, then men, too, could be released from the chore of mowing the backyard by hand.

The tin-can mower

Readers might wonder why I have included the first lightweight domestic rotary lawn mower in my list of 50 world-changing machines. However, not all the machines featured in this book have to be "earth-shattering," like the Model T Ford or the V-2 rocket; there is also room for inventions that changed the world in subtler, gentler ways. The Victa Rotomo lawn mower provides mechanical evidence for an extremely important social phenomenon of the postwar period: the migration of middle-class families from city centers to suburbs. Appropriately, the first Victa lawn mower was built in the garage of an Australian suburban home.

Although he had only completed elementary school, Mervyn Richardson (1893–1972) made his first fortune in the 1920s in auto manufacture and sales, only to lose it during the Great Depression. By 1941, he had fought his way back, working as an engineering salesman, and built his family home in the Sydney suburb of Concord. What every suburban home has, of course, is a front and a backyard, but the domestic lawn mowers on sale in the late 1940s were heavy, cumbersome, and inefficient.

"Richardson noticed that existing lawn mowers were heavy, clumsy machines that were fuel- and energy-inefficient, could not cut long grass or reach a fence line and were difficult to operate."

"BREAKTHROUGH FOR LAWN MOWING" (2009) BY E. GENOCCHIO

The new lightweight mower took suburban Australia and then the world by storm.

THE VICTA ROTOMO FAN MOWER

[A] Throttle and steel handlebars
[B] Fuel tank
[c] Fan
[D] Villiers 98 cc engine
[E] Tin wheels
[F] Steel cover for rotor blade

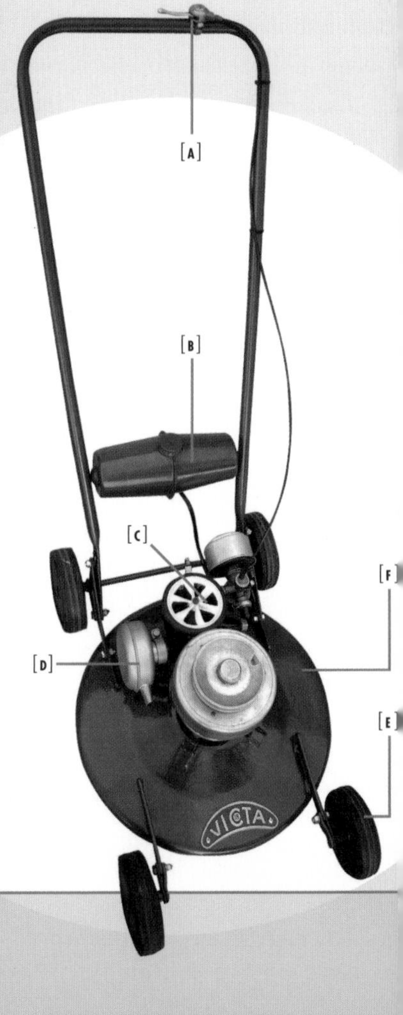

[A]

[B]

[c]

[F]

[D]

[E]

Richardson got interested in mowers when his son started a mowing business to earn cash during his college vacations. He designed and built a couple of mowers to help him out, but even after his son had graduated, Richardson continued to tinker with mower designs in his spare time. In 1952, he showed his family the first Victa (from his middle name, Victor) prototype motorized rotary mower, assembled from a Villiers two-stroke gas engine, mounted on its side turning a rotor, with an empty tin can as the fuel tank. The lash-up mower outperformed anything available at the time. It was light enough for one person to operate and cut the grass perfectly.

Tin wheels with rubber tires for easy traction on the lawn.

Two-stroke gas engine was small, light, and powerful.

A vintage Victa Rotomo before restoration.

Richardson started building the mowers in his garage and placed advertisements in the local paper to announce demonstrations. The response was overwhelming: On demo days, the streets of Concord were congested with the cars of husbands and fathers who did not want to spend their weekends tending to their yards. Within a year, such was the demand that Richardson had quit his job and was working full time manufacturing the new mower. In 1958, Victa Mowers Pty. Ltd. opened its first factory in Milperra, with an annual production of 143,000 mowers destined for sale in 28 countries. The model shown here dates from 1954, and is still close to the original tin-can prototype, though now made of custom parts. The handlebars, frame, and base plate are made of steel mounted on four tin wheels with rubber tires that allowed the Victa to tackle most terrains. A Villiers 98 cc two-stroke gas engine provided the power for the Rotomo, with the gas tank suspended between the handlebars out of harm's way. The addition of a fan provided cooling, which prevented the engine from overheating and stalling. The rotor featured a pivoting blade.

KEY FEATURE:
THE PIVOTING ROTOR BLADE

The Victa Rotomo was fitted with pivoting blades so that if the operator mowed over a rock, the blades would pivot out of harm's way. This ensured a more reliable, consistent cut and prevented damage to the mower or blades. The pivoting action also reduced the vibration reaching the handle, making the whole experience much smoother and more comfortable.

The pivoting rotor blades gave a smoother cut.

38

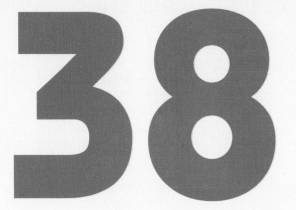

Designer:
UKAEA

MAGNOX NUCLEAR REACTOR

Manufacturer:

UKAEA

Industry
Agriculture
Media
Transport
Science
Computing
Energy
Home

1956

When the first Magnox nuclear reactor began generating electricity at Calder Hall, in England, there were high hopes for the future of nuclear power, but after decades of high-profile accidents, culminating in the Fukushima tragedy of 2011, the future of nuclear power has been questioned by many in the developed world. However, the past decade's relentless rise in oil prices and the poor performance of renewable technologies have led several governments to reconsider their decisions to end or scale down their nuclear power programs.

The dawn of the "Atomic Age"

The decade after the end of the Second World War presents the most striking contrasts. For the first time in history, the entire world had experienced 6 years of total war, pitting the Allies—the U.S., UK and USSR—against the Axis powers—Germany, Italy, and Japan. The newsreels of the time show the waves of joyous hysteria that swept the capitals of the world in 1945 on Victory in Europe Day (May 8) and Victory in the Pacific Day (August 15). The victory, however, had been won at the cost of around 60 million lives, including an estimated 150,000–250,000 casualties (in August and September) from the atomic bombings of Japan. The Atomic Age had been born with the dreadful devastation of Hiroshima and Nagasaki, and presaged a terrible outcome for World War Three—once the Soviet Union had tested its own A-bomb in 1949, and the fragile alliance between the U.S. and USSR had made way for the 40-year chill of the Cold War.

At the same time as the major world powers were developing their terrifying arsenals of nuclear bombers, submarines, and ballistic missiles, they were also developing the civilian uses of nuclear power. In 1953, President Eisenhower (1890–1969) pledged his country's "determination to help solve the fearful atomic dilemma" in his "Atoms for Peace" speech to the General Assembly of the United Nations. Albert Einstein (1879–1955) and Robert J. Oppenheimer (1904–1967), who had encouraged the U.S. to develop the bomb before the Germans, now campaigned vociferously but futilely for the atomic genie to be stuffed back into its bottle. But others saw in nuclear energy a source of clean energy that would transform the world, giving us near endless supplies of cheap electricity, and powering everything from our cars to our domestic boilers and vacuum cleaners. It was in this confused climate that the first commercial nuclear power plant went on line in 1956, at Calder Hall, near the village of Sellafield, in Cumbria, in the northwest of England.

Shattering the atom

The atomic theory of matter dates, like so many other things, from ancient Greece, but it was only with the discovery of the chemical elements, beginning in the eighteenth century, that humans began to unlock the modern understanding of matter. It would take another century for physicists to discover that the atom was made up of smaller particles: the electron orbiting around what was believed to be a solid nucleus. Finally, in the first decades of the twentieth century the nucleus itself was shown to be composed of protons and neutrons. The intriguing possibility then proposed was what would happen if we could split the atomic nucleus. Although in 1905 Einstein had demonstrated in his world-famous equation $E=mc^2$ that "very small amounts of mass may be converted into a very large amount of energy and vice versa"; as late as 1932, he asserted: "There is not the slightest indication that nuclear energy will ever be obtainable. It would mean that the atom would have to be shattered at will."

An illustration of Calder Hall's Magnox reactor, promising a bright new nuclear future.

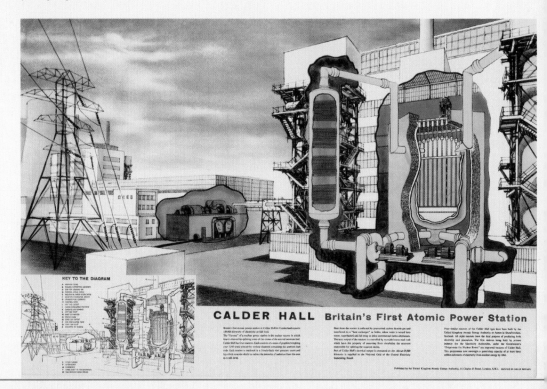

CALDER HALL Britain's First Atomic Power Station

"To the making of these fateful decisions, the United States pledges before you—and therefore before the world—its determination to help solve the fearful atomic dilemma—to devote its entire heart and mind to find the way by which the miraculous inventiveness of man shall not be dedicated to his death, but consecrated to his life." PRESIDENT DWIGHT D. EISENHOWER, "ATOMS FOR PEACE" SPEECH, 1953

Six years later, the great man was to be proved comprehensively wrong when two German physicists, Otto Hahn (1879–1968) and Friedrich Strassmann (1902–1980) bombarded a uranium atom with neutrons, providing the first experimental confirmation of nuclear fission. The pair realized that fission could be maintained by chain reaction, in which the splitting of atomic nuclei produced more neutrons that continued the process until the whole of the uranium was exhausted. A controlled reaction could be used as a power source, as in a nuclear reactor, but an explosive reaction would release enormous amounts of light, heat, and kinetic energy. With war about to break out, the race was on to develop reactors that would be able to enrich uranium to make the weapons-grade uranium and plutonium needed to make a nuclear bomb.

Peaceful atoms

The first nuclear reactors in the U.S. produced the nuclear materials needed to make the first atomic bombs, and military needs continued to drive the programs of the U.S., England, France, and the USSR for the first decade of the Atomic Age. The first experimental nuclear power plant built for civilian use was the AM-1 reactor at Obninsk, 62 miles (100 km) southwest of Moscow, which started generating electricity in 1954; however, Russia did not build another civilian reactor for another decade. In the U.S., the 1950s saw the development of military nuclear applications, including the launch of the SS *Nautilus*, the first nuclear-powered submarine in 1955.

NUCLEAR REACTORS

Chicago Pile-1	1942
Hanford reactors	1943
EBR1	1951
Obinsk reactor	1954
Magnox reactor	1956

MAGNOX NUCLEAR REACTOR

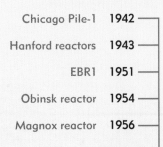

Part of the cooling system at the
Sizewell nuclear power plant in England.

In October 1956, Queen Elizabeth II
(b. 1926) opened the world's first large-scale
commercial nuclear power plant. The UKAEA's
(UK Atomic Energy Authority; 1954–2004)
Calder Hall power station initially operated one
Magnox reactor, producing 60 Mw of electricity), later increased to four reactors producing
200–240 Mw. For the first 8 years of its opera-
tional life, reactor No. 1 at Calder Hall served
the dual purpose of producing electricity for
the national grid and plutonium for Britain's
nuclear weapons program. In total 11 Magnox
power plants were built in the UK, and a further
two were exported to Japan and Italy. Calder
Hall was decommissioned in 2003, having
operated continuously for 47 years without
major mishap.

THE MAGNOX NUCLEAR REACTOR

KEY FEATURE:
THE MAGNOX FUEL RODS

The unenriched uranium fuel rods
were clad in Magnox, an alloy
composed of magnesium, with
aluminum and other metals. The
name is derived from magnesium
non-oxidizing. Although it has the
advantage of a low neutron capture
cross-section, by limiting the
temperature of the core, it reduces
the thermal efficiency of the
reactor, and because the alloy is
reactive with water, Magnox fuel
rods cannot be stored in water for
extended periods.

The Magnox reactor had a simple design, which was an advantage from the point of view of safety. Like other nuclear reactors, the Magnox was powered by uranium, but while later designs would use enriched uranium (containing 2–3 percent of uranium-235), Magnox used natural uranium (containing 0.7 percent uranium-235). The reactor core was encased in a pressurized steel or concrete containment vessel. In order to keep the nuclear reaction under control and slow the neutrons, the reactor had a graphite moderator core. The Magnox-alloy-clad fuel rods were inserted into vertical channels that ran through the core. Boron control rods could be inserted to absorb the neutrons and halt the nuclear chain reaction. Using natural uranium meant that the fuel elements had to be changed more often, but the reactor was designed to allow refueling without a complete shut down. Once the reactor was turned on, and the fission chain reaction began, the core reached extremely high temperatures. The reactor needed to be cooled to avoid a catastrophic core fire or meltdown. In the Magnox the coolant was carbon dioxide gas. The hot gas transferred its heat to water through a heat exchanger, converting it to steam, which drove turbines that generated electricity.

The bulbous cooling towers of a nuclear power plant.

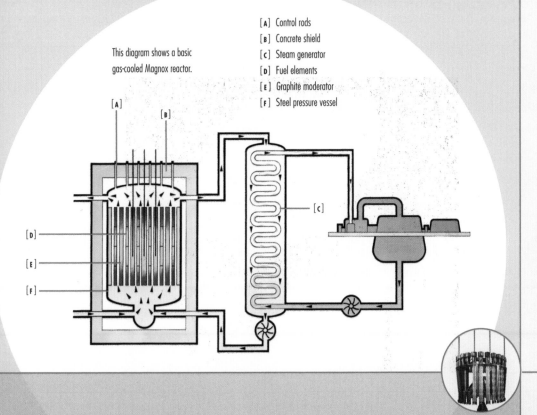

This diagram shows a basic gas-cooled Magnox reactor.

[A] Control rods
[B] Concrete shield
[C] Steam generator
[D] Fuel elements
[E] Graphite moderator
[F] Steel pressure vessel

39

UNIMATE 1900

Industry ■

Agriculture

Media

Manufacturer:

Transport

Unimation Inc.

Science

Computing

Energy

Home

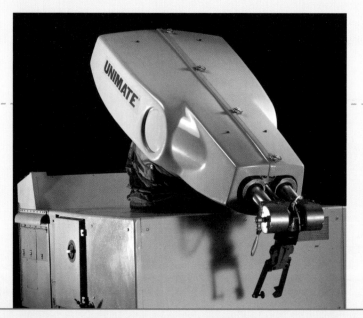

1961

The installation of the Unimate industrial robot at a G.M. diecast plant in New Jersey marks the beginning of large-scale industrial automation, and the subsequent transformation of factory work in the developed world. Although inspired by Isaac Asimov's robot books, the Unimate is nothing like the walking, talking humanoid robot portrayed in popular science fiction.

I, Unimate

In 1966, Johnny Carson (1925–2005) hosted a very special guest on *The Tonight Show*: a series 1900 Unimate robot. To the delight and amusement of Carson, his crew, and the studio audience, and no doubt to the considerable relief of its Unimation operators, the robot successfully putted a golf ball, opened a beer can and poured it into a mug, and conducted the show's band. Admittedly, there were a few cheats: The beer, for example, had to be partly frozen, because the Unimate's "hand" lacked the sensitivity not to squeeze the can and spray its contents over the studio, and the "conducting" seemed fairly wooden. Nevertheless, the performance (available to watch online) was an incredible marketing coup for the manufacturer Unimation, its founder, the inventor, George Devol (1912–2011), and president, Joseph Engelberger (b. 1925).

"In 1961 we got our opportunity to put our innovation to the test at G.M.'s diecasting plant [....] We were concerned about how the diecast machine operators would react to this man replacement. In fact, their consensus was that our machine was a curiosity destined to fail."

G. MUNSON, "THE RISE AND FALL OF UNIMATION INC."
FROM *ROBOT* (2010)

ROBOTS

Televox	1926
Gakutensoku	1928
Elektro	1937
Elmer and Elsie	1948
Unimate	1961

UNIMATE 1900

In 1961, Unimation had installed the world's first industrial robot at the General Motors (G.M.) plant in Ewing Township, NJ. Devol and Engelberger were initially worried that the G.M. workers would reject the robot and try to block its introduction. However, unlike nineteenth-century textile workers, who had smashed up mechanized looms, the G.M. workforce was unconcerned about the possible robotic competition; they were sure that the Unimate was doomed to fail. Diecasting was an ideal choice for robotic automation: It was a dirty, dangerous, and repetitive job. The Unimate 001 picked up red-hot car parts that had just been diecast, dropped them into coolant, and passed them to the assembly line, eliminating any need for human handling.

In 1969, two developments assured the future of Unimation, and established industrial robots as the cutting edge of industrial technology: G.M. automated its plant in Lordstown, OH, which produced 110 cars an hour, twice as many as any other plant; and the Japanese firm Kawasaki licensed Unimation technology and started to manufacture and sell industrial robots in Japan and East Asia.

Unimate robots on an automated
auto production line.

THE UNIMATE 1900

KEY FEATURE:
PROGRAMMABILITY

Engelberger demonstrated the programmability of the Unimate robot on *The Tonight Show* in 1966. Using a remote console plugged into the robot's arm, he entered the sequence of movements that the robot had to "learn" and reproduce. In this case, he programmed the robot to conduct an orchestra with a baton.

Programming a Unimate for a new task with the control pad.

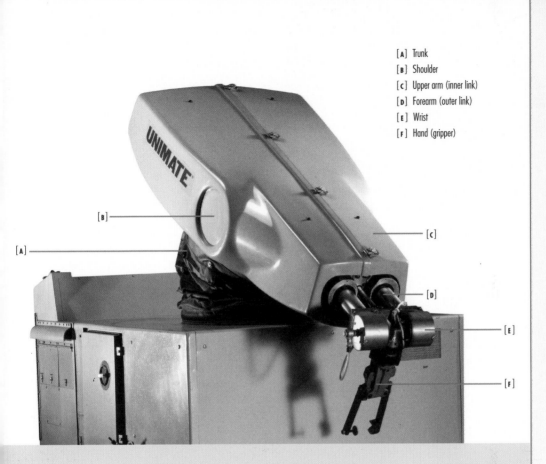

[A] Trunk
[B] Shoulder
[C] Upper arm (inner link)
[D] Forearm (outer link)
[E] Wrist
[F] Hand (gripper)

UNIMATE

Although inspired by the robots of sci-fi novelist Isaac Asimov (1920–1992), the Unimate 1900 was not Robbie from *Lost in Space*. Although humanoid robots such as Televox and Elektro had been created in the 1920s and 30s, these were more fairground attractions than working industrial robots. The Unimate was basically a programmable articulated arm, with step-by-step commands stored on a magnetic drum. For their prototype, Devol and his design team had opted for a hydraulic drive that would have a smoother action than an electric motor, but the hydraulics of the day were primitive and leaked. The arm, mounted on a large stand could rotate around its "trunk," and up and down at the "shoulder." The upper arm held the extensible "forearm" but was not articulated at the "elbow" like later models. The "wrist" rotated and could be fitted with different types of "hands" or grippers. The Unimate's dexterity was first demonstrated in public in 1961 at a trade show in Chicago, where it was programmed to pick up letters to spell out simple phrases. Although Devol got interest from the Ford Motor Company, their first customer was General Motors.

40

Designer:

Wernher von Braun

SATURN V
ROCKET

Manufacturer:

NASA

Industry
Agriculture
Media
Transport
Science
Computing
Energy
Home

1967

As we saw in entry 33, the German V-2 formed the basis for the space programs of the U.S. and USSR. Within 16 years, both the Russians and Americans would put a man into space. American pride, stung by early Soviet successes in the space race, set itself a truly impressive (and expensive) goal: to get man to the moon by 1969. The project would require the largest launch vehicle ever constructed, the Saturn V.

The past meets the future

On December 20, 1968, the crew of *Apollo 8* sat down to lunch on the eve of their departure. Their mission was not one of the six moon landings between 1969 and 1972, but their historic spaceflight would be the first to take humans out of Earth's orbit and around the moon. We can only imagine that the atmosphere in the crew dining room that day at Cape Kennedy, FL, must have been fairly tense. Between 1961 and 1968, there had been many failures of unmanned rockets but amazingly only one human fatality: Soviet cosmonaut Vladimir Komarov (1927–1967). Normally the crew would not be allowed guests before a mission because of the danger of infection, but possibly because of the momentous nature of their mission, as well as the identity of the visitor, an exception had been made in this case.

The unexpected lunch guest was Charles Lindberg (1902–1974), the first man to fly solo, nonstop across the Atlantic. In view of the safety record of early aircraft, Lindberg's crossing in 1927 was probably more hazardous than the first lunar orbital flight. Over lunch, Lindberg told the *Apollo* crew of his meeting with Robert Goddard (1882–1945) in the 1930s, when the father of rocketry had envisaged flights to the moon, which he conceded might cost as much as $1 million. The actual cost of the Apollo program was many orders of magnitude higher. In 1966 alone, NASA's budget peaked at $4.5 billion, or 0.5 percent of U.S. GDP at the time. As lunch drew to a close, Lindberg asked his hosts how much fuel they would be using for liftoff. One of the crew made a quick calculation and answered 20 tons/sec. The aviator observed, "In the first second of your flight tomorrow, you'll burn ten times more fuel than I did all the way to Paris."

Winning the space race

Eight years before the Lindberg lunch, there was no Apollo program—not even an American rocket capable of getting an unmanned probe to the moon. All the Americans seemed to be able to do was to play catch-up with the Soviets. The Russians had been the first to launch a satellite into Earth's orbit with *Sputnik 1* in 1957, and they had followed that coup with the first orbit of the Earth by Yuri Gagarin (1934–1968)

aboard *Vostok 1* in April 1961. It was an extremely piqued President John F. Kennedy who addressed a joint session of Congress a month later. To restore American pride, he had to come up with something big, and he did not disappoint: He announced that the U.S. would land a man on the moon before the end of the decade. Although true to his word, as the first moon landing took place in 1969, Kennedy had fallen to an assassin's bullet 6 years before.

The largest man-made object ever sent into space blasts off from the Kennedy Space Center.

The man who was chosen to realize the president's vision was German rocket engineer Wernher von Braun (1912–1977), who had designed the V-2 rockets during the Second World War. His Nazi past meant that he had been kept out of the limelight between 1945 and 1957, but because of the Soviet successes in space and the U.S. Navy's repeated failures with its Vanguard rockets, the government turned to von Braun, who was developing the Army's Jupiter rocket. Having caught up with the Russians thanks to von Braun, NASA was ready to embark on the greatest scientific adventure of the twentieth century, the Apollo program. In 1960, von Braun was appointed director of the Marshall Space Flight Center in Huntsville, AL, where development of the Saturn program began in earnest. In 1963, after considering several different alternatives for the moon mission, including multiple launchers and the assembly of the spacecraft in Earth's orbit, NASA opted for a single launch vehicle to take the command and lunar modules to the moon: the C-5 rocket, renamed the Saturn V.

Wernher von Braun posing next to the Saturn's massive S-1-C engines.

"First, I believe that this nation should commit itself to achieving the goal, before this decade is out, of landing a man on the moon and returning him safely to the earth. No single space project in this period will be more impressive to mankind, or more important in the long-range exploration of space; and none will be so difficult or expensive to accomplish."

PRESIDENT JOHN F. KENNEDY (1917–1963),
FROM A SPEECH TO CONGRESS, MAY 1961

MANNED SPACECRAFT

Vostok 1	**1961**
Mercury	**1961**
Mercury MA 6	**1962**
Vostok 6	**1963**
X-15	**1963**
Soyuz 1	**1967**
Saturn V	**1967**

SATURN V ROCKET

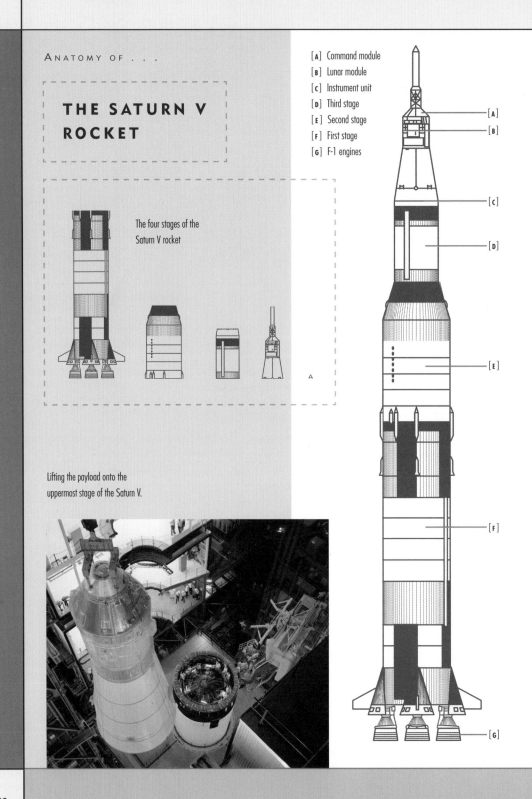

ANATOMY OF . . .

THE SATURN V ROCKET

The four stages of the
Saturn V rocket

Lifting the payload onto the
uppermost stage of the Saturn V.

[A]   Command module
[B]   Lunar module
[C]   Instrument unit
[D]   Third stage
[E]   Second stage
[F]   First stage
[G]   F-1 engines

[A]
[B]
[C]
[D]
[E]
[F]
[G]

The third stage of the Saturn V rocket used in the *Apollo 7* flight.

A description of the Saturn V will quickly exhaust any writer's store of superlatives. Let's begin with a comparison that might make sense to a layperson. The largest jetliner, the Airbus 380-800, is 262 ft (80 m) long by 23 ft (7 m) wide, compared to the Saturn's 363 ft (111 m) height (with the *Apollo* spacecraft attached) and 33 ft (10 m) diameter. Granted the A-380 can carry 519 passengers, and the Saturn V, a crew of three, the A-380 has a maximum range of 9,500 miles (15,400 km), which would only get you ¹/₂₃ of the way to the moon. The Saturn V consisted of three stages (S-IC, SII, and S-IVB) with their own engines; the instrument unit; and the payload. Like the V-2, the three stages used liquid oxygen as the oxidizer, but the first stage used refined petroleum (RP-1) as fuel, while stages 2 and 3 used liquid hydrogen (LH2). S-IC's five F-1 engines delivered 34 meganewtons of thrust to propel the rocket to an altitude of 220,000 ft (67 km) in 168 seconds; S-II had five J-2 engines providing 5.1 meganewtons of thrust to take the rocket through the upper atmosphere; and S-IVB had a single J-2 engine that was the only engine that could be started twice during a lunar mission. The instrument unit sat on top of the whole assembly and controlled the rocket from liftoff to S-IVB separation.

KEY FEATURE:
PAYLOAD

Although it's often said that "size doesn't matter," in rocketry, it definitely does. With a low-Earth-orbit payload capacity of 3,306 tons (3,000 metric tons) and a lunar payload of 90,000 lb (41,000 kg), it was the only launch vehicle large enough to get the *Apollo* spacecraft to the moon. In 1973, the last Saturn V, modified to a two-stage rocket, launched Skylab into Earth's orbit.

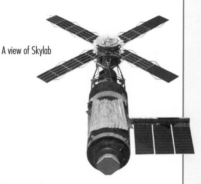

A view of Skylab

Saturn V reaches a massive acceleration at blastoff.

41

Designer:

Godfrey Housefield

EMI CT SCANNER

Manufacturer:

EMI

Industry

Agriculture

Media

Transport

Science

Computing

Energy

Home

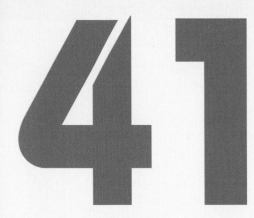

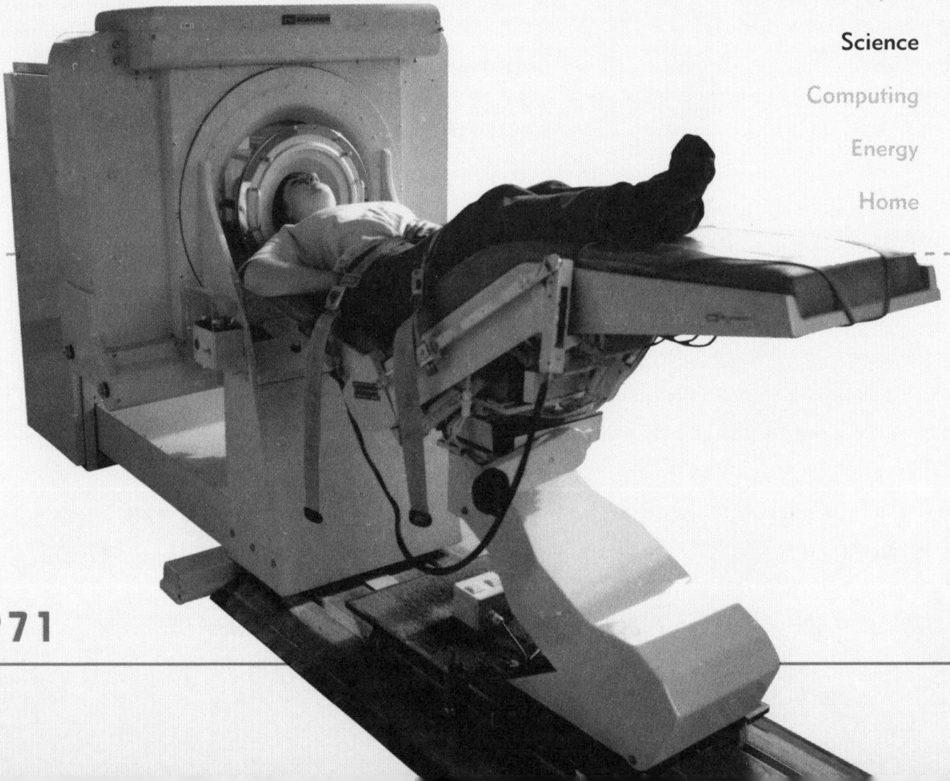

1971

The discovery of x-rays in the late nineteenth century revolutionized medical diagnostics. However, x-rays had certain imaging limitations that were not resolved until the introduction of computed tomography by EMI in 1971. The subsequent development of the CT scanner has led to 3-D imaging of the internal structures of the human body.

The Beatles and the brain scanner

EMI is best known for its role in the music business, however, between the 1940s and 1980s, the firm had an electronics manufacturing division making radar equipment during the war and broadcasting hardware in the postwar period. In 1958, EMI developed the first transistorized computer in the UK under the leadership of Godfrey Housefield (1919–2004). Backed by the money that EMI was making after its signing of the Beatles in 1962, Housefield began work on a revolutionary new medical imaging system, x-ray computed tomography, now known simply as CAT or CT scanning.

In 1967, Housefield visited the leading neurological facility of the day in the UK, the National Neurological Hospital in London, to propose the construction of a new type of x-ray scanner that would produce images of slices of the brains of patients. The head of neuroradiology replied that the existing techniques—pneumoencephalography, plane tomography and angiography—were already able to provide imaging for diagnosis, and he could not see any use for a new type of scanner. In reality, the techniques he listed were far inferior to what Housefield was proposing. Pneumoencephalography, for example, entailed draining most of the cerebral fluid and replacing it with a gas to improve the quality of brain x-rays. The procedure, however, was extremely painful and dangerous and required a 2- to 3-month recovery period.

Undeterred by the curt refusal, Housefield arranged a meeting with the head of neuroradiology at the Atkinson Morley Hospital (AMH) in southwest London, where he received a much better reception. Within 4 years, the CT scanner had produced the first brain images at AMH and overnight revolutionized the field of neurology. Although the earliest scans had a relatively low resolution (80 x 80 pixels), they provided an unequalled diagnostic tool that was described at the time as "being better than a roomful of neurologists." CT, which was being developed independently in the U.S., was later able to provide 3-D images of the brain and other anatomical structures.

THE EMI CT SCANNER

[A] X-ray tube
[B] Detectors
[C] Rotating gantry
[D] Couch for patient

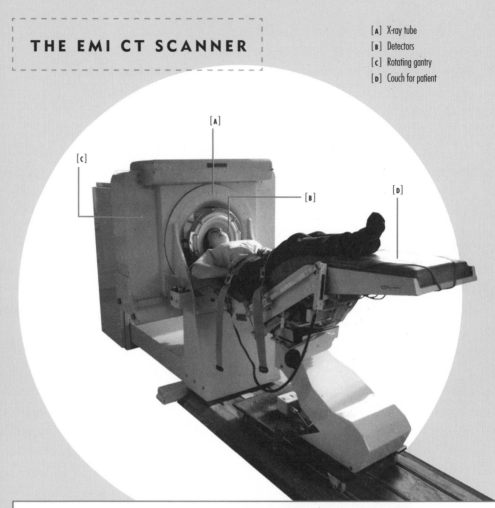

[A]

[C]

[B]

[D]

KEY FEATURE:
ALGEBRAIC RECONSTRUCTION TECHNIQUE

Algebraic Reconstruction Technique (ART) is an iterative algorithm that is used for the reconstruction of images from a series of angular projections. The technique was adapted for use in computed tomography by Godfrey Housefield for the 1971 CT scanner.

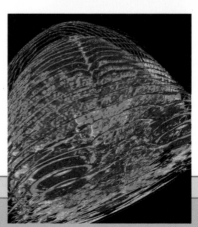

A representation of the brain and eyes produced by a CT scanner.

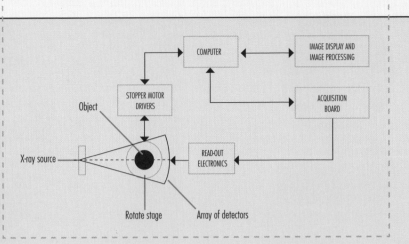

	COMPUTER	← →	IMAGE DISPLAY AND IMAGE PROCESSING	A schematic representation of a modern CT scanner.

STOPPER MOTOR DRIVERS

ACQUISITION BOARD

Object

X-ray source

READ-OUT ELECTRONICS

Rotate stage   Array of detectors

The CT scanner is a development of the x-ray tomography technique discovered at the beginning of the twentieth century, in which a radiologist made a sectional image through the patient's body by moving an x-ray tube and the plate film in opposite directions during exposure. The first prototype scanner at AMH produced its first scan on October 1, 1971. The original machine was only designed to take brain scans, hence only the patient's head was placed within the device. In 1973, Housefield would go on to develop a whole-body scanner. The scanner had an x-ray tube mounted on the top of the mobile gantry with a single photomulti-plier detector on the bottom. When in operation the scanner took 160 parallel images through 180 degrees, with each scan taking a little over 5 minutes. The data was sent on tape to a large computer to be processed using Algebraic Reconstruction Technique (ART), which took 2.5 hours. The first commercial EMI scanner, acquired the images in about 4 minutes, and the computation time per picture was 7 minutes. The commercial version of the scanner required a water-filled Plexiglass tank and a rubber head-cap around the patient's head to reduce the range of the x-rays reaching the detectors.

"Hundreds of radiologists, neurologists and neurosurgeons from around the world headed for Wimbledon to see the new machine at AMH. Orders poured in to EMI despite the then astonishing $300,000 price tag."

THE INTERNET JOURNAL OF NEUROSURGERY (2010) BY A. FILLER

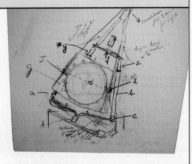

A preliminary sketch by the creator of the CT scanner, Geoffrey Housefield.

42

Designer:
Shizuo Takano

JVC
HR-3300EK

Manufacturer:

JVC

Industry

Agriculture

Media ■

Transport

Science

Computing

Energy

Home

1976

In the mid-1970s, recording sound and images on tape was not a new idea. Reel-to-reel video technology for professional use was developed immediately after reel-to-reel tape recorders. The real contest was between rival video formats. By 1976, the competition had been whittled down to two major players: Sony's Betamax and Matsushita-JVC's VHS.

All-out war

The greatest "war" of the second half of the 1970s was not a pseudo-colonial tussle between the superpowers. The Vietnam War had ended in 1975, and the U.S., USSR, and Communist China were settling down to a period of relatively peaceful, if rather chilly, coexistence. The parties squaring off in this conflict of the titans were two giant Japanese electronics firms: Sony and Matsushita (now Panasonic), the parent company of JVC, and the battleground was the lucrative home VCR market. Video recording was not a new technology; Ampex had developed reel-to-reel video recorders in the 1950s, and in 1970, the Dutch electronics firm Philips launched its N1500 videocassette format, hoping to repeat its 1963 success with compact audiocassettes. However, Philips had not quite got it right. Additionally, in the mid-70s, the Japanese dominated the world's electronics industry, and both consumers and industry analysts waited to see which of the Japanese formats would become the world standard.

"The format war between Betamax and VHS became a 'winner takes all' situation [....] The media industry had seen format competition before, such as in recorded music between Edison's waxed cylinder and Berliner's disc in the first part of the century and between 45s and LPs more recently [....] Video formats, on the other hand, had great trouble coexisting."

Veni, Vidi, Video (2001) by F. Wasser

JVC HR-3300EK

A Sony Betamax machine (below) and a size comparison of the Betamax and VHS cassettes (left).

THE JVC HR-3300EK

KEY FEATURE:
THE VHS CASSETTE

The VHS cassette was a plastic shell 7 x 4 x 1 in (19 x 10 x 2.5 cm), with a flip cover that protected the tape when it was out of the machine. In the original VHS format, the audio was recorded on a linear track at the upper edge of the tape. The HR-3300EK could be used to record sound only, but the AUDIO DUB would also engage the video REC function, though the video would record and playback a blank screen.

A front view of a VHS cassette

A rear view of a VHS cassette

Having failed to reach an agreement over a Japanese standard, Sony launched Betamax in 1975, hoping to establish an unbeatable lead over rival JVC who released VHS ("Video Home System") in 1976 in Asia and Europe and in 1977 in the U.S. At the height of the war, hostilities spread to the two companies' customers, who each defended their choice of VCR. Supporters of Betamax argued that not only was the tape smaller and easier to store, it also had much better sound and picture quality. The VHS side countered that while Betamax tapes were just 60 minutes, VHS were 120 minutes—later extended to 240 minutes (which was coincidentally just long enough to tape the NFL Super Bowl). And even if their VCRs produced slightly less clear pictures and sound, they cost a lot less than the Sony machine. In the end, extended tape duration and low cost proved to be an unbeatable combination with the consumer.

The operating keys of the HR-3300EK should be familiar even to the iPod generation: "PLAY," "STOP," "REW," "FF," "REC," "PAUSE," and "EJECT" are self explanatory, but the machine also had an "AUDIO DUB" key for sound recording only. The keys locked mechanically so that if you were playing a video, you had to press "STOP," before you could select "REW" or "FF." A row of eight buttons selected the channel to be viewed or recorded. Three switches below the channel button were the mode, input and output selectors for the machines. All three switches had to be set in the correct position to ensure that you could record, play, or watch TV. Fortunately for the confused operator, this function was automated on later models. The digital clock and timer controls were on the bottom left. Upon depressing the EJECT key, the cassette loader rose vertically to allow removal or insertion of the cassette. Once the PLAY key was depressed, the machine pulled the tape from the cassette and wrapped it around the head drum, which rotated at 1,800 rpm (NTSC) or 1,500 rpm (PAL). The tape loading action was known as "M-lacing" as the tape was drawn out by the threading posts and wrapped around half of the head drum, giving it the approximate shape of the letter M.

[A] Operating keys [F] Hours/minutes
[B] Counter [G] Cassette loader
[C] Channel selectors [H] Main function selector
[D] Clock [I] Output selector
[E] Timer control [J] Record selector

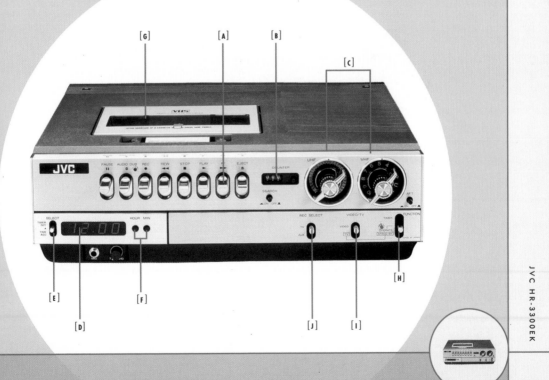

43

Designer:
Jay Miner

ATARI
2600

Manufacturer:

Atari Inc.

Industry

Agriculture

Media

Transport

Science

Computing

Energy

Home ■

"Be a flying ace, a race car champion, a tennis star and a space pioneer all in one afternoon with the Video Computer System by Atari, the new computerized electronic video game system for home TV that is designed to give you the most sophisticated, intricate and fun video games." ATARI PRESS ADVERTISEMENT, 1977

1977

The Atari 2600 was not the first games console, but it was the one that established home video gaming for a generation of children and teens. With its introduction, childhood games and toys were changed forever, creating the "Screen Generation," who preferred to stay indoors to interact with their TVs and computers over playing outdoors or going to video arcades.

Killer App

I am from a generation whose toys were mostly made of inanimate wood, metal, and plastic and only "interactive" if you had a powerful imagination. As we got older, we got electric toys: battery-powered cars, train sets, and, my personal favorite, the Scalextric race car track. But as I and my friends reached our teens and began to hang out in games arcades—then still mainly populated by pinball machines—there arrived a new kind of machine: the video arcade game, starting with very simple games such as the table tennis game PONG in 1972. It was not long before machines of increasing sophistication took over, leading to one of several attempts to convert the arcade formats into consoles for home use. But as these early consoles could only play one game, they were usually well behind the latest arcade hit and sold poorly.

The R&D team at Atari, which included chip designer Jay Miner (1932–1994), decided to build a multi-game platform that would offer the user maximum flexibility. The VCS (Video Computer System; later renamed the Atari 2600), released in 1977, provided the most advanced audio and graphics of the time. It sold well, through the Sears department store chain in the U.S., but what it needed was a "killer app"—a game that would make it the must-have toy of the decade. In 1978, Japanese games designer Tomohiro Nishikado (b. 1944) designed a game that was going to take the world of gaming by storm: "Space Invaders." Inspired by the 1898 *War of the Worlds*, the game pitted ranks of descending tentacled aliens against the player's laser cannon. When Atari released the VCS version of the game in 1980, they had found their killer app. The Atari 2600 went on to outsell all its rivals, and its huge success bred complacency that would be one of the factors that caused the North American video game crash of 1983.

ATARI 2600

THE ATARI 2600

[A] On/off
[B] TV signal selector
[c] Difficulty (player A)
[D] Games cartridge
[E] Difficulty (player B)
[F] Game select
[G] Reset

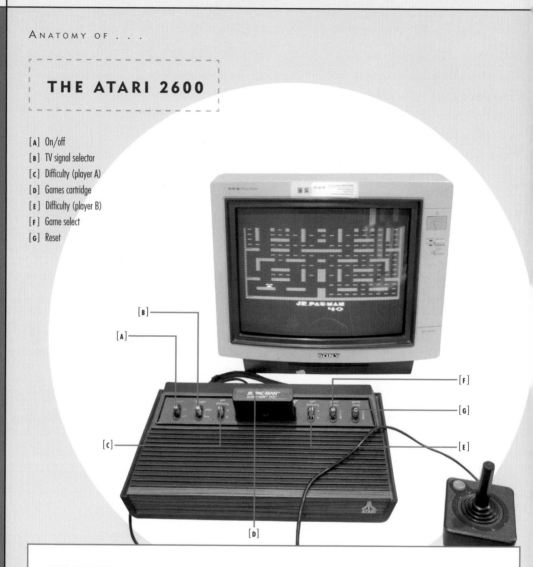

KEY FEATURE: THE TELEVISION INTERFACE ADAPTER (TIA)

The most striking difference between the Atari and modern PCs and consoles is in working memory: The Atari made do with a pitifully low 128 bytes of RAM. At that time, the cost of memory was such that any more would have made the console too expensive. Miner's solution was to do away with the frame buffer altogether. Instead his TIA generated the playfield and five separate graphic objects on every scan line from their respective registers. The single-color playfield consisted of a register 20 bits wide that could be mirrored to the other half of the screen, to make it 40 bits wide, with a palette of 128 colors. The five objects were "player A" and "player B" (two single-color 8-pixel horizontal lines); two single-color "missiles," horizontal lines varying in width from one to 8 pixels; and a "ball," a horizontal line the same color as the playfield.

The motherboard of the Vader 2600, which was identical to the 2600-A motherboard.

In the earliest models, the console had a row of six switches along the top, split into two groups of three by the cartridge port. From left to right: power; b/w or color TV; difficulty player A; difficulty player B; game select; and reset. The two difficulty switches were later moved to the rear, along with the two ports for the input devices that came with the console: two joysticks and two paddles, and the TV port. The 2600 was packaged with ten games. Unlike earlier consoles in which the games were stored on internal chips, the Atari system stored games on ROM chips in the cartridges themselves.

"Paddle" controllers for the Atari 2600 console.

Atari consoles were available in a number of styles, including this "wood veneer" model.

44

Designer:

Kozo Ohsone

SONY TPS-L2 "WALKMAN"

Manufacturer:

Sony Corporation

Industry

Agriculture

Media ▪

Transport

Science

Computing

Energy

Home

1979

The Sony TPS-L2 "Walkman" cannot claim to be the first portable music platform, but it was the one that successfully marketed the concepts of portability and individual playlist customization to the world. Its huge commercial success confirmed Japan's place as the world leader in the field of consumer electronics for the next two decades.

The invention that never was

In 1972, Andreas Pavel (b. 1945) came up with a revolutionary concept in personal entertainment: a portable audiocassette player that he called the "Stereobelt." The what? I hear you ask. Exactly. Having failed to interest several major electronics firms, because they felt that consumers would not want to be seen in public wearing headphones (!), Pavel applied for patent protection in 1978. But in 1979, before his patents had been granted, Sony released the TPS-L2, which, after a few early naming mishaps, became known under the hugely successful global brand name of the Sony "Walkman."

Pavel not unnaturally sued Sony, and the fight turned into a regular David vs Goliath fight, but not one settled with a slingshot but with something much deadlier: attorneys at law. The epic court battle lasted a quarter of a century. In 2004, Sony and Pavel finally settled out of court for an undisclosed sum rumored to be in the region of $10 million. So, Pavel got his recognition, some cash, and a happy ending, although 32 years too late, and well after the audio cassette Walkman had been relegated to the outdated-technology museum shelf.

"The Walkman was initially launched as 'Soundabout' in the U.S., 'Stowaway' in England, and 'Freestyle' in Australia. However, the name 'Walkman' was eventually accepted overseas, as Walkman portable stereos became very popular in Japan and tourists visiting Japan from abroad started buying them as a souvenir."

SONY PRESS RELEASE, 1999

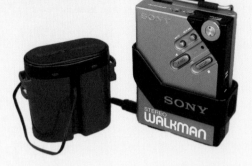

The silver-and-black Sony Walkman, with belt clip and the plastic carrying case for the extra batteries this power-hungry cassette player required.

By the 1970s, the personal stereo (PS) was an invention that was waiting to happen. There was nothing particularly new about the technology: In 1963, Philips had launched the audiocassette and small tape recorder/player with built-in loudspeaker, and headphones were as old as radio. But as Pavel discovered, for most electronics executives, listening to music meant putting on your favorite Beethoven LP on the hi-fi in the mahogany cabinet in the living room, and not running around the streets listening to a PS strapped to your belt through headphones. What they failed to grasp was that teens, who had been laboriously assembling personalized cassette playlists for a decade, wanted a player that would allow them to take their music out of the bedroom. The three men who recognized this desire and its huge marketing potential were the septuagenarian founder of Sony, Masaru Ibuka (1908–1997), its 58-year-old chairman, Akio Morita (b. 1921), and 46-year-old Tape Division Manager, Kozo Ohsone (b. 1933).

THE SONY TPS-L2 "WALKMAN"

KEY FEATURE:
THE "WALKMAN" CONCEPT

The Sony Walkman was above all a brilliant marketing concept that both met and created the need for personal stereos with user-made playlists. Targeted at teens, the new "Walk-men" and "Walk-women," it confirmed Japan's domination of consumer electronics. In 1986, what had sounded in 1979 like a strange Japanese-English hybrid word — "Walkman" — was included in the Oxford English Dictionary.

The solar-powered Walkman (1987), a member of the large and growing Walkman family.

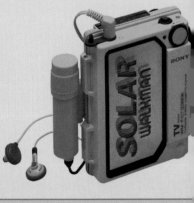

[A] Two headphone jacks "Guys & Dolls"
[B] Hotline button (Orange button)
[C] FF/REW
[D] Play
[E] Stop/Eject
[F] L/R volume controls
[G] Metal case
[H] Headphones

When Ibuka, Morita, and Ohsone decided to go ahead with the Walkman—meeting, it is said, some considerable resistance within Sony—they had a ready-made small cassette recorder that they could adapt: the TCM-600 "Pressman," designed for use by journalists to tape interviews. Ohsone's team stripped out as much as they could from the metal casing to reduce the size, weight, and cost— the recording circuitry, record and pause keys, mic socket, erase head, loudspeaker and tape counter—and fitted a stereo head, and two sliding volume controls on the side beneath the main control keys (PLAY, FF, REW, STOP/EJECT).

The Walkman had two headphone jacks (though it came packaged with only one set of MDR-3L2 headphones), labeled "Guys & Dolls." Beside the jacks was the orange "hotline" button that faded the music and mixed in the output of a small internal mic, so that two listeners could hear each other or a third party without stopping the tape. Although originally designed for speech rather than music, the Walkman's hardware was considered to give extremely good sound reproduction. The trademark blue-and-silver case was sold with a plastic battery case and a belt clip.

45

Designer:

Henrik Stiesdal

VESTAS
HVK10

Manufacturer:

Vestas

Industry

Agriculture

Media

Transport

Science

Computing

Energy

Home

1979

While governments and energy companies were pouring billions into nuclear fission and fusion research and prospecting for dwindling reserves of crude oil, Danish grass-roots "self-builders" were designing machines to harvest one of nature's freely available and renewable energy resources: wind power. Henrik Stiesdal's Vestas HVK15 was one of several wind turbines that were developed commercially in Denmark in 1979.

Blowing in the wind

In 1978, two men stood in a field in Denmark proudly watching something that looked like a collision between a propeller plane and an electricity pylon. The duo were Henrik Stiesdal and Karl Erik Jørgensen (d. 1982), two of many wind-power enthusiasts who were designing and building wind turbines to generate electricity for their homes in rural Denmark in the mid-70s. It was Jørgensen's second turbine, and his colleague and later business partner in Herborg Vindkraft (HVK), Henrik Stiesdal, then still a college student, persuaded him to add a third blade to his original two-blade design. The turbine used a fiberglass rotor custom-made by Økær, which itself was a

year-old startup. Although further technical difficulties had to be overcome, including a complete redesign of the air-braking system, within a year the fledgling HVK had a turbine capable of producing 30 kilowatts.

Generating energy from the wind is one of the most ancient power-generating technologies. Wind-powered devices are known from antiquity, and the first windmills are attributed to ninth-century Iran. The idea of generating electricity with windmills dates back to the late nineteenth century. But in the ages of coal-powered steam and oil-powered internal combustion, there was little need to develop renewable energy sources such as solar and wind power.

"The technology was difficult, but no more difficult than a science-minded college student could manage. It was complex in that it brought together many different technologies, from generator systems to gearbox control systems to towers and so on."

HENRIK STIESDAL (B. 1957)

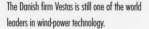

The Danish firm Vestas is still one of the world leaders in wind-power technology.

By the late 1970s, however, the world had experienced the 1973 oil shock and was becoming disillusioned with nuclear power. The Danish self-builders were not driven to create wind turbines purely by environmental concerns, but also to provide cheap energy for their families and businesses in a country with few other national energy resources than a plentiful supply of wind. Having perfected their first commercial turbine, HVK signed a licensing agreement with the agricultural equipment manufacturer Vestas to produce the HVK10 in 1979. Along with other Danish wind-power startups Kuriant, Nordtank, and Bonus, Vestas launched the "Danish Concept" in wind power.

THE VESTAS HVK 10

KEY FEATURE:
THE BLADE-TIP AIR BRAKES

In 1978, the HVK prototype and another Danish turbine fitted with Økær blades both experienced catastrophic failures in high winds. Although the turbines had internal hub brakes, in both cases these failed and both turbines suffered serious damage. In response, HVK and Økær designed the first air brakes fitted to the blade tips. Initial models employed an external design; however, it proved too noisy and was later replaced by an internal air brake.

Wind turbines are typically located on hilltops, facing into prevailing wind.

The Vestas HVK 10 was a horizontal-axis wind turbine (HAWT) with three 16-ft (5-m) fiberglass Økær 5 blades (33-ft/10-m rotor) generating 30 Kw of electricity; a smaller model with 10-ft (3-m) Aerostar blades (19.6 ft/6 m rotor) delivered 22 Kw. Unlike later designs, which were mounted on solid towers, the early Vestas used a metal lattice pylon as the support for the turbine. For maximum efficiency, a turbine needs to be able to turn to face an oncoming wind. In smaller turbines, this is achieved with a weather vane but larger models require a separate yaw drive controlled by a weather vane and anemometer (device measuring wind speed). The wind flowing over the blades turns the rotor attached to the main low-speed shaft. The shaft activates the gearbox, which spins the generator producing the electrical output. Wind turbines can be grouped together in wind farms that are connected to a country or state's electricity infrastructure, or, as was intended for the initial "Danish Concept," to serve a single dwelling, farm, or workshop.

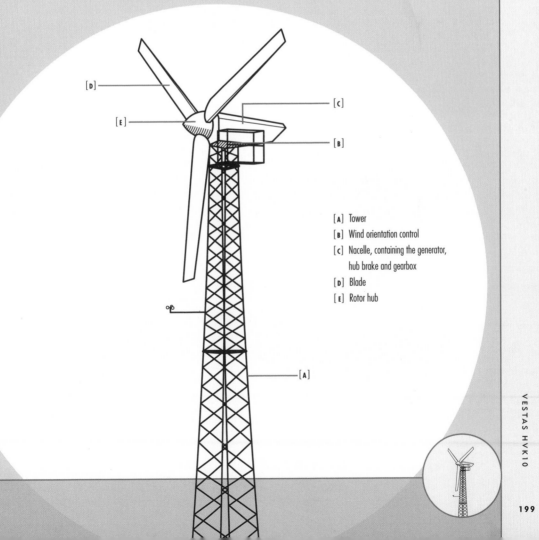

[A] Tower
[B] Wind orientation control
[C] Nacelle, containing the generator, hub brake and gearbox
[D] Blade
[E] Rotor hub

46

Designer:
Don Estridge

IBM PC
5150

Manufacturer:

IBM

Industry

Agriculture

Media

Transport

Science

Computing ■

Energy

Home

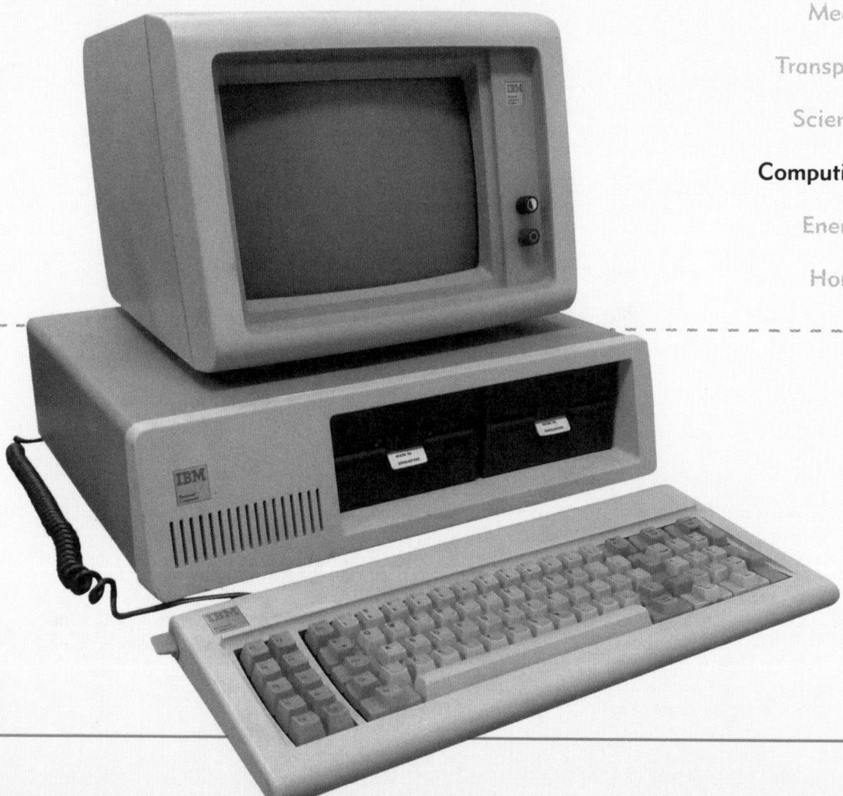

1981

It would be difficult to overstate the impact of the introduction in the late '70s and early '80s of the first generation of home PCs on the development of human society and culture. The IBM PC 5150 and its many clones first transformed the world of work and education, and they were then brought into the home as an all-purpose workstation and entertainment center. In the 1990s, the ubiquitous PC desktops and laptops would connect us all to the World Wide Web.

From mainframe to bedroom

Like other college students in the late 1970s, I had ostensible access to a computer, although it was not in my dorm room, or even in the library. It was the computer department's room-sized mainframe, and to use it I'd have to book time on a terminal in the computer room. Using the computer was not done lightly, as you'd have to know how to program it. And without any handy word-processing apps, you wouldn't have dreamed of using it for something as complex as word processing. For that, you had a typewriter: electric—if you were lucky (and rich)—or clunky, manual—if you were me. Amazingly (for the benefit of the iPad generation), we managed to lead fairly normal, happy lives without 2-GB HDs.

We first encountered the concept of the "computer" with Babbage's Difference Engine. A giant mechanical hand-cranked calculator for polynomial functions, the Engine, nevertheless had many of the features of much later machines, including basic programmability and a printer. In 1936, British mathematical genius

Alan Turing (1912–1954) established the basis of modern computer science. Although driven to suicide because of persecution over his sexuality, Turing was later recognized in the description of all modern computers as "Turing Machines." During WWII Turing worked on the UK's code-breaking Colossus computer, while independent computer development took place in Germany (Z3) and the U.S. (ENIAC), whose building-sized military number crunchers consisted of tens of thousands of vacuum tubes and relays.

The transistor was invented in 1947, and by 1955 had replaced vacuum tubes in computers. The original input was with punch cards or punch tape, which was later replaced by magnetic tape. Although by the '50s computers were getting smaller, faster, and cheaper—the IBM 650 weighed a mere 2,000 lb (900 kg)—we were still a very long way from a portable home PC. The 1960s witnessed the development of the microprocessor, and in 1971, Intel produced its first 4-bit CPU (central processing unit), the Intel 4004.

IBM PC 5150

Holy Trinity

Although the first portable, stand-alone computers went on sale in the early 1970s, these were often sold in small numbers to enthusiasts in kit form. The age of the mass-market microcomputer began with three machines, later dubbed the "Trinity," which were released a few months apart in 1977: the Commodore PET (Personal Electronic Transactor), the Apple II, and the Tandy RadioShack TRS-80. The PET was the least successful of the three because of its small calculator-like keyboard; the Apple II already showed the distinctive Macintosh styling and featured a full QWERTY keyboard and color graphics capability. Although the most expensive of the Trinity, the Apple II was the longest lasting and bestselling of the three. The TRS-80 combined its CPU and keyboard, with a separate monitor and power supply. Although less advanced than the Apple, it had a full QWERTY keyboard, was small, and half the price.

Seeing the success of the Trinity, and the Atari 400/800 released in 1978, the world's largest computer manufacturer at the time, International Business Machines Corp (IBM), decided to get into the microcomputer business. Codenamed "Project Chess," the development of the IBM PC bypassed the company's usual R&D protocols. Because team leader Don Estridge (1937–1985) wanted to go to market as soon as possible, his team sourced existing components from other manufacturers and also raided existing IBM products. Although the keyboard and main unit were original designs, the monitor came from IBM Japan and the printer from Epson. Development took about one year, and IBM released the PC in August 1981, at a price of $1,565 for the basic model. But perhaps the most far-reaching decision was Estridge's choice of the Intel 8088 processor paired with Microsoft's DOS 1.0, over an IBM processor and OS. The enormous success of the IBM PC ensured that MS DOS, developed by Bill Gates (b. 1955), would become the world's dominant operating system.

One of the first Apple home computers
with an external Hayes Micromodem.

The Toshiba 1100, one of many IBM-PC clones to flood the market.

Since the 1880s and the widespread introduction of the typewriter (see page 84), anything connected with typing was the domain of mainly female typists and secretaries. Post IBM PC, however, typists would be replaced by data-entry operatives, and executives would have to learn to type their own letters and reports. Along with the office, the PC penetrated the home, where it transformed the world of gaming, and the schoolroom. Within a decade, it would become the platform to access the World Wide Web.

Attack of the clones

Although not the first computer to be called a "PC," the name quickly became synonymous with the IBM desktop and its many imitators. The first clones (PC-compatible computers) were announced a year later, including the Compaq "Portable." Within a few years, all rival OS and computer architectures would disappear under the PC/MS DOS onslaught, with one notable exception: Apple Inc., which released its first "Macintosh" in 1984. While IBM and American manufacturers congratulated themselves on yet another world-changing innovation, their complacency was about to be shaken: The Japanese were coming. With the genius for anticipating what the consumer both needed and wanted, and packaging it at a reasonable price (as with the Sony Walkman), Toshiba introduced the first mass-market laptop, the T1100 in 1985.

PERSONAL COMPUTERS

Xerox Alto	1973
Altair microcomputer	1974
Commodore PET 2001	1977
Apple II computer	1977
Tandy TRS-80	1977
Atari 400/800	1978
TI-99/4 PC	1979
Sinclair ZX-80	1980
VIC-20	1981
IBM 5150 PC	1981

IBM PC 5150

THE IBM PC 5150

[A] Keyboard
[B] Floppy disk drive A
[C] Floppy disk drive B
[D] Main unit
[E] Screen

"IBM is proud to announce a product you may have a personal interest in. It's a tool that could soon be on your desk, in your home or in your child's schoolroom. It can make a surprising difference in the way you work, learn or otherwise approach the complexities (and some of the pleasures) of living." **IBM** PRESS ADVERTISEMENT, 1981

Apart from the bulky CRT monitor, the IBM PC still looks like something we would recognize as a desktop, with its familiar arrangement of screen, main unit, and keyboard. However, in terms of operation and capabilities, that's pretty much as far as the similarities go. The CPU provided by the Intel 8088 ran Microsoft's v.1.0 DOS—something so far removed from a modern OS that let's not even go there—with a lavish 640 KB of RAM. And what about the HD, you ask? What HD? Naturally we are in a time before optical discs and CDs, so all the PC's apps and work were accommodated on two 5.25-inch (13.3 cm) floppy disk drives that were either 160 KB or 360 KB. In order to work on the IBM PC, you had to play a complex game of

The original IBM-PC keyboard had an annoying built-in clacking to remind the user of a typewriter, but this was quickly phased out.

pass the parcel with the diskettes. In the initial concept, the main storage for the PC was to be provided by an external compact cassette drive; however, as DOS was only sold on disk, the cassette idea was a nonstarter, and in 1983, one of the floppy disk drives was replaced by a 10 MB HD. The PC's keyboard set the industry standard, but its first design had an annoying clacking built in—no doubt to remind users of the comforting sound of typewriter's keys—which was quickly phased out.

KEY FEATURE: THE PC CONCEPT

Although not the first personal computer, and possibly not the most advanced technically at the time, the IBM PC firmly established the PC in the home, school, and workplace. The market dominance of PC clones also ensured the supremacy of the Microsoft operating system, MS DOS and later MS Windows.

The 5150's 5¼" floppy disk drive with a DOS 1.1 disk.

1981 IBM PC motherboard with 16 KB of RAM (expandable to 64 KB).

47

Designer:

Dale Heatherington

HAYES
SMARTMODEM 300

Manufacturer:

Hayes Microcomputer Products

Industry

Agriculture

Media

Transport

Science

Computing ■

Energy

Home

"What, exactly, is the Internet? Basically it is a global network exchanging digitized data in such a way that any computer, anywhere, that is equipped with a device called a 'modem,' can make a noise like a duck choking on a kazoo." DAVE BARRY (B. 1947)

1981

The development of home computers in the late 1970s and of the IBM PC in 1981 created ready-made terminals that could be connected by fixed telephone lines into even more complex computer networks that would one day grow into the WWW. But before we could get the Internet as we know it today, we needed a device to link the thousands of PCs scattered across the world: the modem.

Only connect

Until the 1980s, unless you were working for certain government organizations and universities in the U.S. and Western Europe, you wouldn't have even heard of the different "nets" that connected their computers electronically. But as far back as the early '70s, people were swapping files through FTP, and sending emails, and, from 1978, "spamming" one another. In the late 1980s, I was working in Japan. I recall the awe that I felt when I sent my first-ever email to our New York office—which I probably wrote out in longhand before typing onto the screen. I can't remember exactly what we were using to send the email, but it was going out from a PC clone through a modem (modulator-demodulator) that seemed so fiendishly complicated at the time that we had to have an IT technician set it up and

establish the real-time connection for us. Of course, within a few weeks, we were probably all firing emails left, right, and center.

Compared to today's seamless, automated 30-megabit Wi-Fi router-modems connections, 300-bit modems were incredibly slow and clunky, and you'd often get half the message or file sent, and then the line would go down or the computer would crash, meaning that you'd have to start all over again. When I got my first home connection in the 1990s, I was fortunate that in 1981, Dennis Hayes (b. 1950) and Dale Heatherington (b. 1948) had developed the first fully automated modem, the Smartmodem 300, and the industry-standard Hayes command set that could dial, answer, and hang up the phoneline—and all that without a telephone, but being plugged directly into the phone socket. Early modems used an acoustic system to transmit data over the phone line, and the modem made its trademark sound that Dave Barry compared to "a duck choking on a kazoo."

Dale Heatherington with the prototype
Hayes 80-103 300 bps modem.

48

Designer:

Leroy Hood

ABI 370A DNA SEQUENCER

Manufacturer:

Applied Biosystems Inc.

Industry

Agriculture

Media

Transport

Science ■

Computing

Energy

Home

"The tools we develop are used for a diverse range of applications—to discover genes and proteins that are tied to diseases, to discover polymorphisms that may affect drug safety and efficacy […], to provide early detection of dangerous pathogens, and to provide overwhelming evidence of guilt or innocence in serious crimes." C. BUZRICK QUOTED IN "APPLIED BIOSYSTEMS" (2006) BY M. SPRINGER

1987

The sequencing of the human genome was one of the most ambitious scientific projects initiated in the late twentieth century, on a par with the Apollo moon landings and the construction of the Large Hadron Collider. The mapping of the genome is now transforming preventative medicine, diagnostics and treatment.

Cracking the code

If you want to read about some real code breaking, forget *The Da Vinci Code*—child's play—or the cracking of Germany's wartime "Enigma" codes by Britain's first electronic computer. The Human Genome Project (HGP) began in 1990, with international funding, with the aim of mapping and deciphering the human genome (HG). The HG consists of 23 pairs of chromosomes, carrying 20–25,000 individual genes (and a lot of other stuff), which is made up of 3.3 billion DNA base pairs of the four nucleosides: adenine, guanine, cytosine, and thymine (A, G, C, T). The decoding and sequencing of the genome took 13 years, at a cost of $3 billion.

DNA was first recognized in human cells in the late nineteenth century, and in 1927 geneticists proposed that inheritance might be controlled by a chemical mechanism in the cell's nucleus, but it was only in 1953 that James Watson (b. 1928), Francis Crick (1916–2004), and Rosalind Franklin (1920–1958) created the first accurate model of the DNA double helix. Although the basic structure of the genome was understood, its complexity remained daunting.

The first sequencing methods developed were slow, complex, and involved the use of toxic chemicals and radioactive materials. In 1986, Leroy Hood (b. 1938) at Caltech developed a semi-automated dye-terminator sequencing machine, which used four fluorescent dyes to identify and scan the nucleoside bases.

Licensed to Applied Biosystems, the prototype was developed a year later into the first automated DNA sequencing machine, the ABI 370A. Furnished with the new sequencing technology, the HGP was completed much faster and at a fraction of the cost. The sequencing of an individual's genome, which cost $3 billion in 2003, is estimated to go down to $1,000 by 2014.

Swiss biologist and doctor Johannes Friedrich Meischer (1844–1895) was the first man to isolate nucleic acid.

ABI 370A DNA SEQUENCER

49

Designer:
Lyman Spitzer

HUBBLE SPACE TELESCOPE

Industry
Agriculture
Media
Transport
Science ■
Computing
Energy
Home

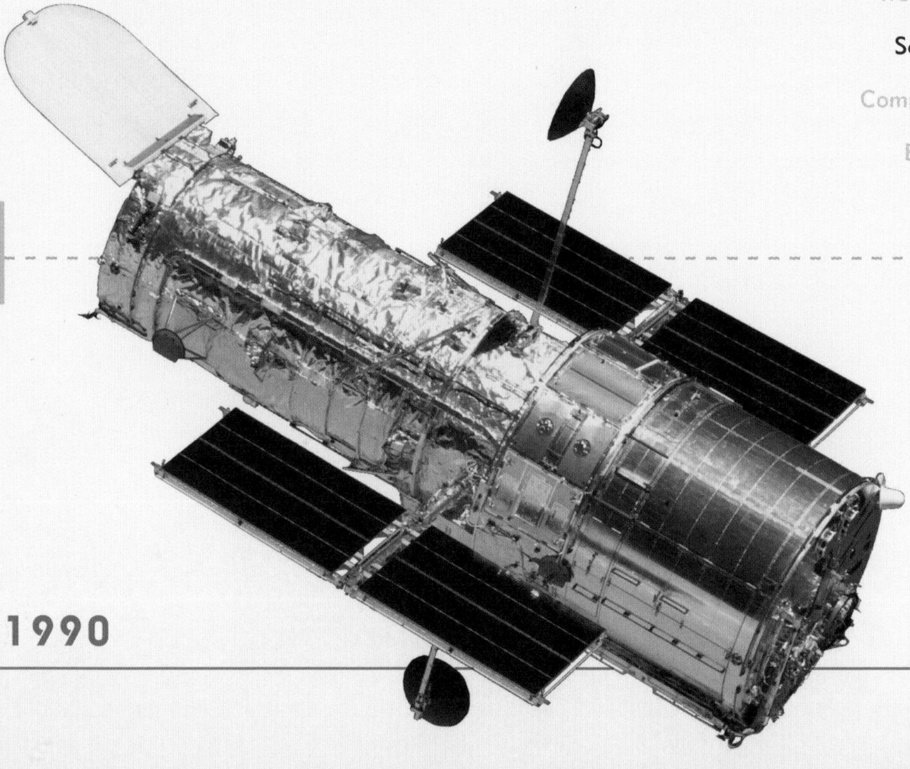

Manufacturer:

PerkinElmer Inc.

1990

Launched into orbit in 1990, the Hubble Space Telescope's mission was compromised by human error during its construction. However, once it had been repaired in 1993, it began to provide the clearest images of extremely faint and distant objects, revealing not only the structure of the contemporary universe but also its distant past, as the farther we look outward from Earth, the further we go back in time.

I can't see clearly now

We've all done it: We've followed the instructions scrupulously, and whatever it is—new cake recipe or set of garage shelves—is going to be your best ever. And then you realize that one very small but incredibly significant measurement at the very beginning was wrong and that the whole thing just won't work. If it's home improvement or baking, you can just start again, but if it's a space telescope orbiting 347 miles (559 km) above your head, you've got a slightly bigger problem. The space shuttle *Discovery* delivered the Hubble Space Telescope (HST) into orbit in April 1990, and within weeks of the launch astronomers realized that the telescope had a serious defect.

SPACE TELESCOPES AND OBSERVATORIES

Orbiting Solar Observatory	1962
International Ultraviolet Explorer	1978
Infrared Astronomical Satellite	1983
Cosmic Background Explorer	1989
Hubble Space Telescope	1990

"We find them smaller and fainter, in constantly increasing numbers, and we know that we are reaching into space, farther and farther, until, with the faintest nebulae that can be detected with the greatest telescopes, we arrive at the frontier of the known universe."

EDWIN HUBBLE (1889–1953)

An analysis of the images revealed that the primary 94 in (2.4 m) mirror was the wrong shape—out by just $^2/_{1000}$ of a millimeter, but it was enough to distort the images coming from very faint distant objects, whose study was one of the Hubble's main objectives.

Astronomer Lyman Spitzer (1914–1997) first proposed a space-based optical telescope in 1946. However, at that time, humans were only beginning to make their first tentative forays out of Earth's atmosphere (see "V-2," page 142). The resolution of Earth-based optical telescopes, he argued, was greatly reduced by Earth's atmosphere, and, of course, seriously impaired by cloud cover. But to realize his vision, Spitzer would have to wait several decades for space technology to advance sufficiently to put a telescope into orbit, and for the funding to be released from the Apollo program. It must have been a bitter blow when he found that his life's work had been ruined by an avoidable human error. However, in 1993, NASA came up with a solution: in the simplest terms, to fit the Hubble with a pair of "eyeglasses" that corrected the distortion.

An image of Saturn taken by the Hubble Space Telescope in March 2004.

THE HUBBLE SPACE TELESCOPE

The Hubble telescope in orbit around Earth.

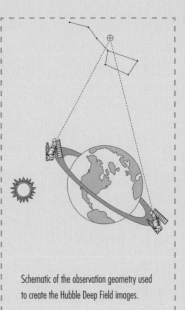

Schematic of the observation geometry used to create the Hubble Deep Field images.

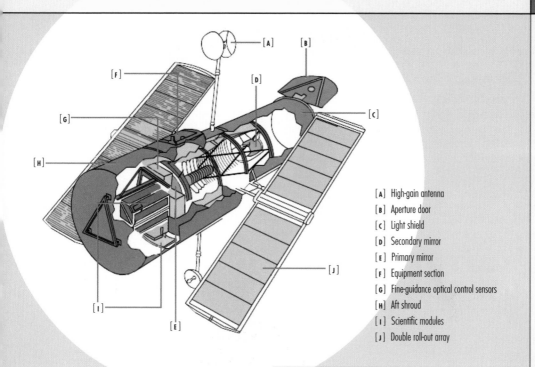

[A] High-gain antenna
[B] Aperture door
[C] Light shield
[D] Secondary mirror
[E] Primary mirror
[F] Equipment section
[G] Fine-guidance optical control sensors
[H] Aft shroud
[I] Scientific modules
[J] Double roll-out array

KEY FEATURE: SPACE

What makes Hubble so much better than any Earth-based telescope is not the size of its mirror (there are much bigger ones on Earth) or the complexity of its electronics, but its location in low Earth orbit outside the distortions caused by the atmosphere. With an orbital period of 97 minutes, the Hubble completes 14–15 full scans of the heavens in one day.

The Hubble Space Telescope is a conventional optical telescope equipped to operate outside of the atmosphere. The design is known as a Ritchey-Chrétien telescope, which is the standard design for large optical telescopes. The 42.3 ft (13 m) long casing houses the 94 in (2.4 m) primary mirror that reflects the light from the object it is observing to a smaller secondary mirror, which focuses the light and sends it to the science package that records the images and sends them back to Earth where they are further processed and enhanced electronically. The original instrument package included five scientific instruments: the Wide Field and Planetary Camera, the Goddard High Resolution Spectrograph, the High Speed Photometer, the Faint Object Camera, and the Faint Object Spectrograph, which operated with visible light and ultraviolet. The UV instruments were far superior to anything comparable on Earth, because the atmosphere filters out most of the UV radiation before it reaches the surface. Power for the HST is provided by two arrays of solar panels on either side of the main casing. But as the Hubble does not have a propulsion unit, it is scheduled to drop out of orbit and crash back to Earth between 2019 and 2032.

50

MOTOROLA
STARTAC

Manufacturer:

Motorola

Industry

Agriculture

Media ■

Transport

Science

Computing

Energy

Home

1996

The last entry is an invention whose impact on every aspect of human civilization continues to grow as its full potential is gradually realized. Although lacking in many of the functions we are now used to on our smart phones, the clamshell Motorola StarTAC, with its basic voice and SMS capabilities, is recognized as one of the breakthrough cell phone model designs—the iPhone of its day.

"Beam me up, Scottie!"

New York City, April 3, 1973, was the time and place of one of the most momentous events of recent media-communication history—on a par with Alexander Graham Bell's first telephone call or John Logie Baird's first television broadcast: the first public demonstration of a call made on a portable cell phone by Martin Cooper (b. 1928), then head of R&D at Motorola. The prototype Motorola DynaTAC that Cooper used to make the call, weighed a hefty 2.2 lb (1 kg) and had a battery life of only 35 minutes, with 20 minutes' talk time. But as Cooper later recalled, "That wasn't really a big problem because you couldn't hold that phone up for that long." It took another decade of research and development before the first U.S. cellular network went live in Chicago, using the now much lighter DynaTAC 8000x.

MOBILE TELEPHONY

First radio call to a car	1906
AT&T Mobile Telephone Service	1947
First car phone	1956
First DynaTAC call	1973
First U.S. cell phone network	1983
First UK mobile phone network	1984
Motorola StarTAC	1996

"We had no idea that in as little as 35 years more than half the people on Earth would have cellular telephones, and they give the phones away to people for nothing."

MARTIN COOPER (B. 1928), INVENTOR OF THE CELL PHONE

MOTOROLA STARTAC

The Motorola DynaTAC 8000X, the granddaddy of all cell phones.

THE MOTOROLA STARTAC

KEY FEATURE:
THE CLAMSHELL DESIGN

Although the 1989 MicroTAC had been announced as the "world's smallest cell phone," which could fit into a shirt pocket, it only had a half flip top covering the keypad, and compared to the StarTAC, it was still relatively large, clunky, and heavy. With the relentless increase in smart phone size and weight, however, the StarTAC clamshell remains one of the most compact phones ever designed.

The first British "mobile phone" service began a year later, and like most of my bemused compatriots, I wondered why anyone would want to walk around talking into a device that looked and handled like a brick and cost several thousand dollars. I held out for over a decade: I made do with an answering machine at home, and, I admit it, a pager. However, in 1996, Motorola, which had been reducing the size and cost of their phones since the original DynaTAC, brought out a new model that was going to transform the cell phone market and convert many holdouts like myself. Building on their success with the 1989 MicroTAC, they released the StarTAC, designed by Rudy Krolopp (b. 1930). One of the undoubted attractions of the StarTAC was its resemblance to the flip-top communicator from the original *Star Trek* series—a similarity that I suspect the manufacturer was playing on when they chose the name and styling.

The first Ericsson cell phone— not so much mobile as portable.

Considering where the cell phone had started in 1973—with the "brick"—the StarTAC is a marvel of miniaturization. It weighed 3.6 oz (102 g) and measured 3.7 x 2 x 1 in (9.3 x 5 x 2.5 cm). Although primitive by the standards of today's smart phones, the StarTAC did the basics: voice calls and SMS text messaging, with small indicator icons for voice mail, SMS, and signal strength on top of the monochrome number/SMS display. The phone memory could hold 99 names/numbers and it saved the last 16 numbers in its caller ID memory.

Opening the phone illuminated the keypad and also answered the phone. The retractable antenna on the side of the phone was used in areas with poor reception. The StarTAC also came with a headphone jack and PC synchronization (later models offered basic Internet access with a small monochrome screen). Along with the clamshell design, the StarTAC was the first phone to have a vibrating alert feature. It came with standard NiMH batteries or optional lithium-ion batteries with a talk time of 210 minutes and standby time of 180 hours.

[A] Loudspeaker
[B] Retractable aerial
[C] Removable lithium-ion battery (at rear)
[D] Number/sms display
[E] Keypad
[F] Microphone

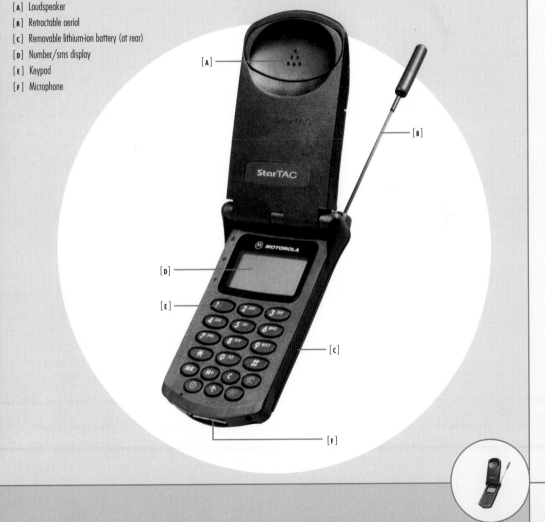

FURTHER READING

Allen, Robert (2009) *The British Industrial Revolution in Global Perspective*, Cambridge: Cambridge University Press

Alphin, Elaine (1997) *Vacuum Cleaners*, Minneapolis, MN: Carolrhoda Books

Balk, Alfred (2005) *The Rise of Radio, from Marconi through the Golden Age*, Jefferson, NC: McFarland & Company

Beaver, Patrick (1987) *The Big Ship: Brunel's Great Eastern*, London: Bibliophile Books

Bodanis, David (2006) *Electric Universe: How Electricity Switched on the Modern World*, London: Abacus

Boothroyd, Jennifer (2011) *From Washboards to Washing Machines: How Homes Have Changed*, Minneapolis, MN: Lerner Classroom

Brandon, Ruth (1996) *Singer and the Sewing Machine: A Capitalist Romance*, New York: Kodansha America

Carlson, W. B. (2003) *Innovation as a Social Process: Elihu Thomson and the Rise of General Electric*, Cambridge: Cambridge University Press

Casey, Robert (2008) *The Model T: A Centennial History*, Baltimore: Johns Hopkins University Press

Casson, Herbert (1910) *The History of the Telephone*, Chicago: A.C. McClurg & Co.

Chaline, Eric (2009) *History's Worst Inventions and the People Who Made Them*, New York: Fall River Press

Chaline, Eric (2011) *History's Worst Predictions and the People Who Made Them*, London: History Press

Chiras, Dan et al (2009) *Power from the Wind: Achieving Energy Independence*, Gabriola Island, BC: New Society Publishers

Cirincione, Joseph (2008) *Bomb Scare: The History and Future of Nuclear Weapons*, New York: Columbia University Press

Collier, Bruce and MacLachlan, James (1999) *Charles Babbage: And the Engines of Perfection*, New York: Oxford University Press USA

Collins, Douglas (1990) *The Story of Kodak*, New York: Harry N. Abrams

Cooke, Stephanie (2009) *In Mortal Hands: A Cautionary History of the Nuclear Age*, New York: Bloomsbury

Cooper, Gail (2002) *Air-Conditioning America: Engineers and the Controlled Environment, 1900-1960*, Baltimore: Johns Hopkins University Press

Croft, William (2006) *Under the Microscope: A Brief History of Microscopy*, Hackensack, NJ: World Scientific Publishing Company

Crump, Thomas (2007) *A Brief History of the Age of Steam: From the First Engine to the Boats and Railways*, Philadelphia, PA: Running Press

Crump, Thomas (2010) *A Brief History of How the Industrial Revolution Changed the World*, London: Robinson Publishing

Edgerton, Gary (2009) *The Columbia History of American Television*, New York: Columbia University Press

Essinger, James (2007) *Jacquard's Web: How a Hand-loom Led to the Birth of the Information Age*, Oxford: Oxford University Press

Ford, Henry (2008) *My Life and Work – An Autobiography of Henry Ford*, Chicago, IL: BN Publishing

Glancey, Jonathan (2008) *The Car: A History of the Automobile*, London: Carlton Books

Gray, Charlotte (2011) *Reluctant Genius: Alexander Graham Bell and the Passion for Invention*, New York: Arcade Publishing

Gustavson, Todd (2009) *Camera: A History of Photography from Daguerreotype to Digital*, New York: Sterling Innovation

Heinberg, Richard (2005) *The Party's Over*, Forest Row, East Sussex: Clairview Books

Henry, John (2008) *The Scientific Revolution and the History of Modern Science*, Basingstoke: Palgrave Macmillan

Herlihy, David (2006) *Bicycle: The History*, New Haven, CT: Yale University Press

Kent, Steven (2008) *The Ultimate History of Video Games: From Pong to Pokemon*, Roseville, CA: Prima Publishing

Kevles, Bettyann (1998) *Naked to the Bone: Medical Imaging in the Twentieth Century*, Reading, MA: Addison-Wesley

Kirby, Richard et al (1991) *Engineering in History*, London: Dover Publications

Loxley, Simon (2006) *Type: The Secret History of Letters*, New York: I.B. Tauris

McCollum, Sean (2011) *The Fascinating, Fantastic, Unusual History of Robots*, Mankato, MN: Capstone Press

McNichol, Tom (2006) *AC/DC: The Savage Tale of the First Standard Wars*, Hoboken, NJ: Jossey-Bass

Millard, Andre (2005) *America on Record: A History of Recorded Sound*, Cambridge: Cambridge University Press

Nowell-Smith, Geoffrey (1999) *The Oxford History of World Cinema*, Oxford: Oxford University Press

Oxdale, Chris (2011)
*The Light Bulb (Tales of
Invention)*, Chicago, IL:
Heinemann Library

Petzold, Charles (2008)
*The Annotated Turing: A
Guided Tour Through Alan
Turing's Historic Paper on
Computability and the Turing
Machine*, Hoboken, NJ:
John Wiley & Sons

Poole, Ian (2006)
*Cellular Communications
Explained: From Basics to 3G*,
Boston, MA: Newnes

Pugh, E.W. (2009)
*Building IBM: Shaping an
Industry and Its Technology*,
Cambridge, MA: MIT Press

Richards, Julia and Scott
Hawley, R. (2010)
*The Human Genome,
Third Edition: A User's
Guide*, Waltham, MA:
Academic Press

Rolt, L. T. C. (2007)
Victorian Engineering, London:
History Press

Sherman, Josepha (2003)
*The History of Personal
Computers*, New York:
Franklin Watts

Smiles, Samuel (2010)
*Lives of the Engineers
George and Robert Stephenson:
The Locomotive*, Charleston,
SC: Nabu Press

Smiles, Samuel (2010)
Men of Invention and Industry,
Charleston, SC: Nabu Press

Sparrow, Giles (2010)
*Hubble: Window on the
Universe*, London: Quercus

Stephenson, Charles (2004)
*Zeppelins: German Airships
1900–40*, Oxford: Osprey
Publications

Stross, Randall (2008) *The
Wizard of Menlo Park: How
Thomas Alva Edison Invented
the Modern World*, New York:
Three Rivers Press

Swedin, Eric and Ferro,
David (2007)
*Computers: The Life Story
of a Technology*, Baltimore,
MD: Johns Hopkins
University Press

Tames, Richard (2009)
Isambard Kingdom Brunel,
Oxford: Shire

Typewriter Topics (2003)
*The Typewriter: An Illustrated
History*, London: Dover
Publications

Von Braun, Wernher (1985)
*Space Travel: An Update
of History of Rocketry and
Space Travel*, New York:
HarperCollins

Walker, Timothy (2000)
*First Jet Airliner: The Story of
the De Havilland Comet Hb*,
Newcastle upon Tyne: Scoval
Publishing

Watson, James (1980)
*The Double Helix: A Personal
Account of the Discovery of the
Structure of DNA*, New York:
Atheneum

USEFUL WEBSITES

Automatic Washer Collectors: *www.automaticwasher.org*

Caterpillar History: *www.caterpillar.com*

Computer History: *www.computerhistory.org*

Early Visual Media Museum: *www.visual-media.be*

General Electric History: *www.ge.com*

Grace's Guide to British Industrial History:
  www.gracesguide.co.uk

Greatest Achievements of the Twentieth Century:
  www.greatachievements.org

Hubble Space Telescope: *www.hubblesite.org*

International Atomic Energy Agency: *www.iaea.org*

Magnox Reactors: *www.magnoxsites.co.uk*

Model T Central: *www.modeltcentral.com*

Motorola Timeline: *www.motorolasolutions.com*

NASA (Apollo Program):
  www.history.nasa.gov/apollo.html

National Geographic History of Photography:
  www.photography.nationalgeographic.com

Robots: *www.robots.com*

Science Museum: *www.sciencemuseum.org.uk*

Smithsonian Institution: *www.si.edu*

Sony History: *www.sony.net*

Total Rewind: History of the VCR: *www.totalrewind.org*

Vintage Appliances: *www.antiqueappliances.com*

Vintage Tools and Gadgets: *www.vintageadbrowser.com*

Wikipedia: *www.en.wikipedia.org*

Wind Power: *www.windsofchange.dk*

IMAGE CREDITS

INDEX